Precision Farming and Geoinformatics

THE AUTHORS

Mr. Vishal D. Wale completed Bachelor's degree from Mahatma Phule Krishi Vidyapeeth, Rahuri (Maharashtra) and Master's degree in Agricultural Meteorology from Mahatma Phule Krishi Vidyapeeth, Rahuri (Maharashtra). He has written number of research paper, popular and technical articles in highly NAAS rated Journals and periodicals. He has 1.5 yrs experience in teaching. Presently he is working as Assistant Professor, College of Agriculture, Maldad, Tal. Sangamner, Dist. Ahmednagar, Maharashtra.

Dr. Pankajkumar V. Patil completed Bachelor's degree from Mahatma Phule Krishi Vidyapeeth, Rahuri (Maharashtra) and Master's degree in Agricultural Entomology from Vasantrao Naik Marathwada Krishi Vidyapeeth, Parbhani (Maharashtra). He did his Docroral degree Ph.D in Agricultural Entomology from Mahatma Phule Krishi Vidyapeeth, Rahuri (Maharashtra.) He was awarded with National Fellowship by UGC, New Delhi during Ph.D. He has 3 yrs experience in teaching and research. He has written 10 research paper, 20 popular and technical articles and short communication in highly NAAS rated and reputed journals and periodicals. Presently he is working as Assistant Professor, College of Agriculture, Maldad, Tal. Sangamner, Dist. Ahmednagar, Maharashtra.

Miss. Kalpana D. Bangar completed Bachelor's degree from Mahatma Phule Krishi Vidyapeeth, Rahuri (Maharashtra) and Master's degree in Agronomy from Vasantrao Naik Marathwada Krishi Vidyapeeth, Parbhani (Maharashtra). She has written number of research paper and technical articles in reputed Journals. Presently she is working as Assistant Professor, College of Agriculture, Maldad, Tal. Sangamner, Dist. Ahmednagar, Maharashtra.

Precision Farming and Geoinformatics

Vishal D. Wale

Pankajkumar V. Patil

Kalpana D. Bangar

2021

Daya Publishing House®

A Division of

Astral International Pvt. Ltd.

New Delhi – 110 002

Publisher's Note:

Every possible effort has been made to ensure that the information contained in this book is accurate at the time of going to press, and the publisher and author cannot accept responsibility for any errors or omissions, however caused. No responsibility for loss or damage occasioned to any person acting, or refraining from action, as a result of the material in this publication can be accepted by the editor, the publisher or the author. The Publisher is not associated with any product or vendor mentioned in the book. The contents of this work are intended to further general scientific research, understanding and discussion only. Readers should consult with a specialist where appropriate.

Every effort has been made to trace the owners of copyright material used in this book, if any. The author and the publisher will be grateful for any omission brought to their notice for acknowledgement in the future editions of the book.

Cataloging in Publication Data--DK

Courtesy: D.K. Agencies (P) Ltd. <docinfo@dkagencies.com>

Wale, Vishal D., author.

Precision farming and geoinformatics / Vishal D. Wale, Pankajkumar V. Patil, Kalpana D. Bangar.

pages cm

Includes bibliographical references and index.

ISBN 978-93-90435-86-9 (HB)

1. Precision farming. 2. Precision farming--Data processing. 3. Geoinformatics. I. Patil, Pankajkumar V., author. II. Bangar, Kalpana D., author. III. Title.

LCC S494.5.P73W35 2021 | DDC 631 23

Published by : **Daya Publishing House®**
A Division of
Astral International Pvt. Ltd.
– ISO 9001:2015 Certified Company –
4736/23, Ansari Road, Darya Ganj
New Delhi-110 002
Ph. 011-43549197, 23278134
E-mail: info@astralint.com
Website: www.astralint.com

Digitally Printed at : **Replika Press Pvt. Ltd.**

Preface

We have a great pleasure in writting the book "*Precision Farming and Geoinformatics*". This book has been written to fulfil the needs of third year sixth semester B.Sc. (Agri.) programme as per syllabus prescribed by the Fifth Deans Committee on Higher Agriculture Education in India. This book covers the information regarding different aspects of precision agriculture, geoinformatics, crop simulation models, remote sensing and nanotechnology. Student need precise information about the subject for preparing for their examination, In a modest way, We have tried to incorporate to the point information of the subject in simple manner.

We are very much thankful to the publisher of this book Mr Anil Mittal for their keep interest and the help during formation of this book cost effective.

We know nobody is perfect in this world; hence suggestions of specific topic will be highly encouraged for improvement of this book. We shall appreciate constructive suggestions from the readers which will improve the quality of the book subsequently.

Vishal D. Wale

Pankajkumar V. Patil

Kalpana D. Bangar

Contents

Chapter 1

Precision Farming

CHAPTER OUTLINES

1.1 Definitions

Precision farming is an information and technology based farm management system to identify, analyze and manage variability within field for farm management. optimum profitability, sustainability and protection of the land resources. **or Precision farming** is the application of technologies and principles to manage spatial and temporary variability associated with all aspects of agricultural production for improving production and environmental quality. **or Precision farming** is the precise application of agricultural inputs with respect to soil, weather and crop need in order to improve productivity, sustainability, quality and profitability in agriculture. **or Precision farming** is the management of agricultural inputs based on the needs of the small individual sites of the fields with the use of variable rate application instead of conventional constant rate of application.

1.2 Concept of Precision Farming

The concept of precision farming entails the use of some high technology equipments for assessing the field conditions and applying chemicals and fertilizers. Managing small area within field to reduce chemical use and improve productivity is the goal of precision farming. Precision farming technology can cover a huge scale of farm land by the support of using **satellite**. Farm field is divided into many small parts and the various data for each part (soil fertility, moisture content, yield, *etc.*) are measured, collected and installed as the database in **GIS (Geographical information system). GPS (Global positioning system)** is used to identify the exact location of both machines and farm field for giving the suitable treatment and operation to meet the condition obtained as the database in GIS. These technologic of precision farming useful for following points;

a) Used to make better decision about many aspects of crop production
b) To maximize the effectiveness of crop inputs
c) To manage and distribute inputs on a site specific basis to maximize long term benefit

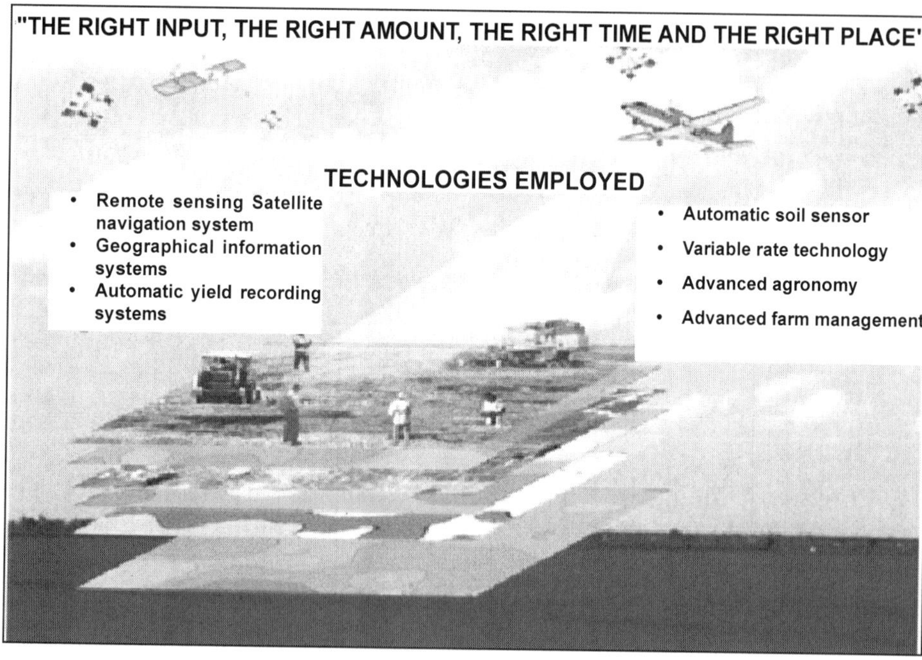

Figure 1.1. Concept of Precision Farming.

d) To manage variability at the sub field level to best utilize resources
e) To achieve a sustainable environmental friendly agriculture
f) Reduction in the human processing task thus reducing the time and efforts
g) Tool to apply fertilizer according to the need of a particular sub field
h) Tool to fight the challenge of agricultural sustainability

It is a modern agriculture practice involving the use of technology in agriculture like remote sensing, GPS and GIS for improving productivity and profitability. It enables farmers to use crop inputs more efficiently including pesticides, fertilizers, tillage and irrigation water. More effective utilization of inputs will bring in more crop yield and quality without polluting the environment and will result in sustainable agriculture and sustainable development. Hence, precision agriculture is about doing the right thing, in the right place, in the right way, at the right time. Pierre Robert is often regarded as the father of precision farming because of his active promotion of the idea and organization of the first workshop, "Soil Specific Crop Management," during the early 1990s. Precision agriculture is a modern farming management concept using digital techniques to monitor and optimise agricultural production processes. Smart farming is the application of information and data technologies for optimising complex farming systems. Smart farming involves not just individual machines but all farm operations. Digital farming is integrating both concepts - precision farming and smart farming.

Many different names *i.e.* precision agriculture, site specific farming, variable rate farming, prescription farming are used for precision farming. The precision farming technologies can be classified into three main groups as data collection technologies, data processing and decision making technologies and variable rate application technologies.

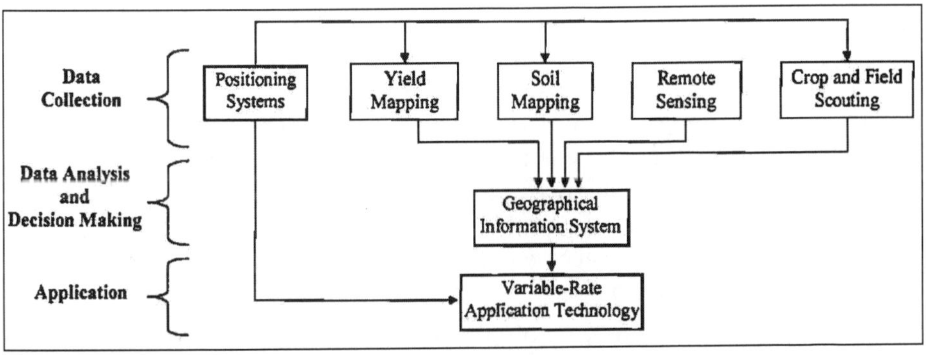

Figure 1.2. Classification of Precision Farming Technologies.

1.3 Need of Precision Farming

a) **For assessing and managing field variability:** Field conditions is difficult to maintain because of the variable farm size, management practices, soil properties, environmental characteristics and changes in areas formed due to annual shifts in leasing arrangements. Precision agriculture offers the potential to automate and simplify the collection and analysis of information.

b) **For doing the right thing in the right place at the right time:** After assessing the variability precision agriculture allows management decisions to be made and implemented in right time in right places on small areas within larger fields.

c) **For higher productivity:** Since precision farming, proposes to prescribe tailor made management practices, it will definitely increase the yield per unit of land, provided nature's other uncontrollable factors are in favor.

d) **For increasing the effectiveness of inputs:** Increased productivity per unit of input used indicates increased efficiency of the inputs.

e) **For maximum use of minimum land unit:** After knowing the land status, a farmer tries to improve each and every part of land and uses it for the production purpose.

1.4 Steps in Precision Farming

1.4.1 Identification and Assessment of Variability

a. **Grid soil sampling:** Grid soil sampling uses the same principles of soil sampling but increases the intensity of sampling compared to the traditional sampling. Soil samples collected in a systematic grid also have location information that allows the data to be mapped. The goal of grid soil sampling is to generate a map of nutrient requirement, called an application map.

b. **Crop scouting:** It is monitoring, inspection, surveillance of crop for general plant health *i.e.* insect, diseases, nutrition and other disorders. **or** It is the process of precisely assessing pest pressure and crop performance to evaluate economic risk from pest infestations and disease, as well as to determine the potential effectiveness of pest and disease control interventions. It involves gathering and recording information about a particular crop on a regular basis to enhance the process of making informed crop management decisions and helpful later when explaining variations in yield maps.

c. **Use of precision technologies for assessing variability:** Faster and in real time assessment of variability is possible only through advanced tools of precision agriculture.

1.4.2 Management of Variability

a. **Variable rate application:** Grid soil samples are analyzed in the laboratory and an interpretation of crop input (nutrient) needs is made for each soil sample. Then the input application map is plotted using the entire set of soil samples. The input application map is loaded into a computer mounted on a variable rate input applicator. The computer uses the input application map and a GPS receiver to direct a product delivery controller that changes the amount or kind of input (fertilizer), according to the application map.

b **Yield monitoring and mapping:** Yield measurements are essential for making sound management decisions. However, soil, landscape and other environmental factors should also be weighed when interpreting a yield map. Used properly, yield information provides important feedback in determining the effects of managed inputs such as fertilizer amendments, seed, pesticides and cultural practices including tillage and irrigation. Since yield measurements from a single year may be heavily influenced by weather, it is always advisable to examine yield data of several years including data from extreme weather years that helps in pinpointing whether the observed yields are due to management or climate induced.

c. **Quantifying on farm variability:** Every farm presents a unique management puzzle. Not all the tools described above will help determine the causes of variability in a field and it would be cost prohibitive to implement all of them immediately. An incremental approach is a wiser strategy, using one or two of the tools at a time and carefully evaluating the results and then proceeding further.

1.4.3 Evaluation of Precision Farming

a. **Economic analysis:** Reduce agrochemicals use, higher nutrient use efficiencies, increased efficiency of management and increased productions of soils from degradation are the benefits to the environment. Enabling technologies can make precision farming feasible, agronomic principles and decision rules can make it applicable and enhance production efficiency can make it profitable.

b. **Environmental assessment:** The reduced use of agrochemicals and inputs being used at right time, in right proportion and right place increased efficiency of their utilization and improve soil, water and ecological sustainability of the agriculture system.

c **Transfer of technology:** The term technology transfer could imply the precision agriculture occurs when individuals or firms acquire and use the technology. The key term to manage is to improve the application of technologies to manage spatial and temporary variability. The communication of precision farming with the farmer associated with the managerial capability of the operator, the spatial distribution of infrastructure and the compatibility of the technology to individual farms will change radically as precision farming continues to develop.

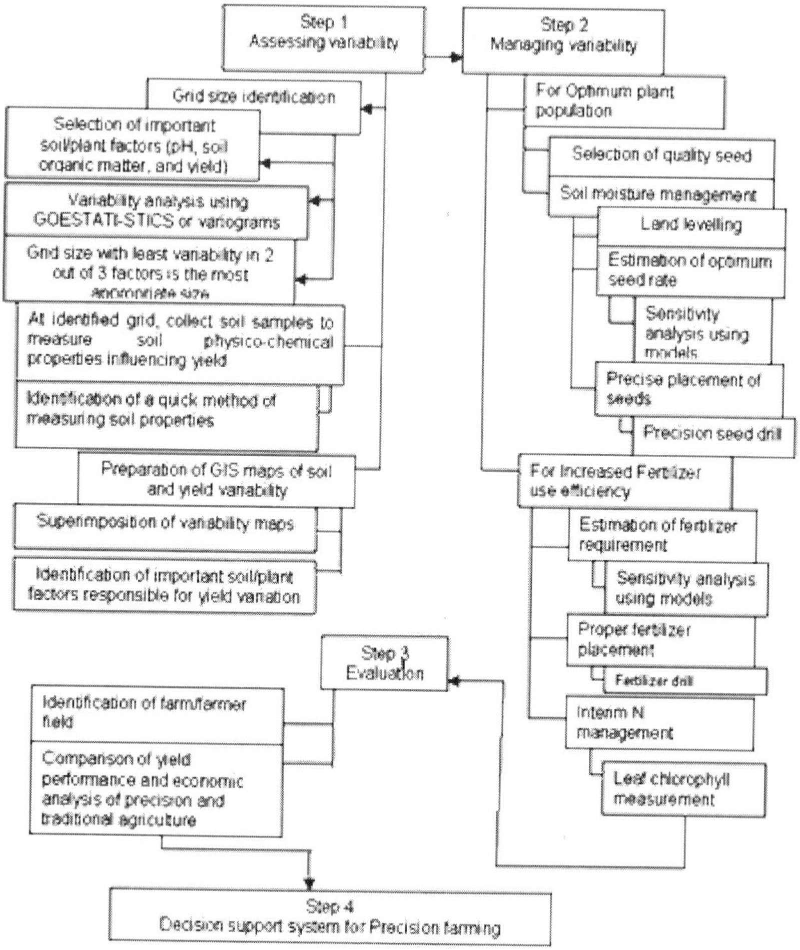

Figure 1.3. Steps in Precision Farming.

1.5 Techniques of Precision Farming

a) Global positioning system
b) Differential global positioning system
c) Sensor technologies
d) Remote sensing
e) Geographic information system
f) Variable rate technologies
g) Rate controllers
h) Variable rate application
i) Software
j) Yield monitor

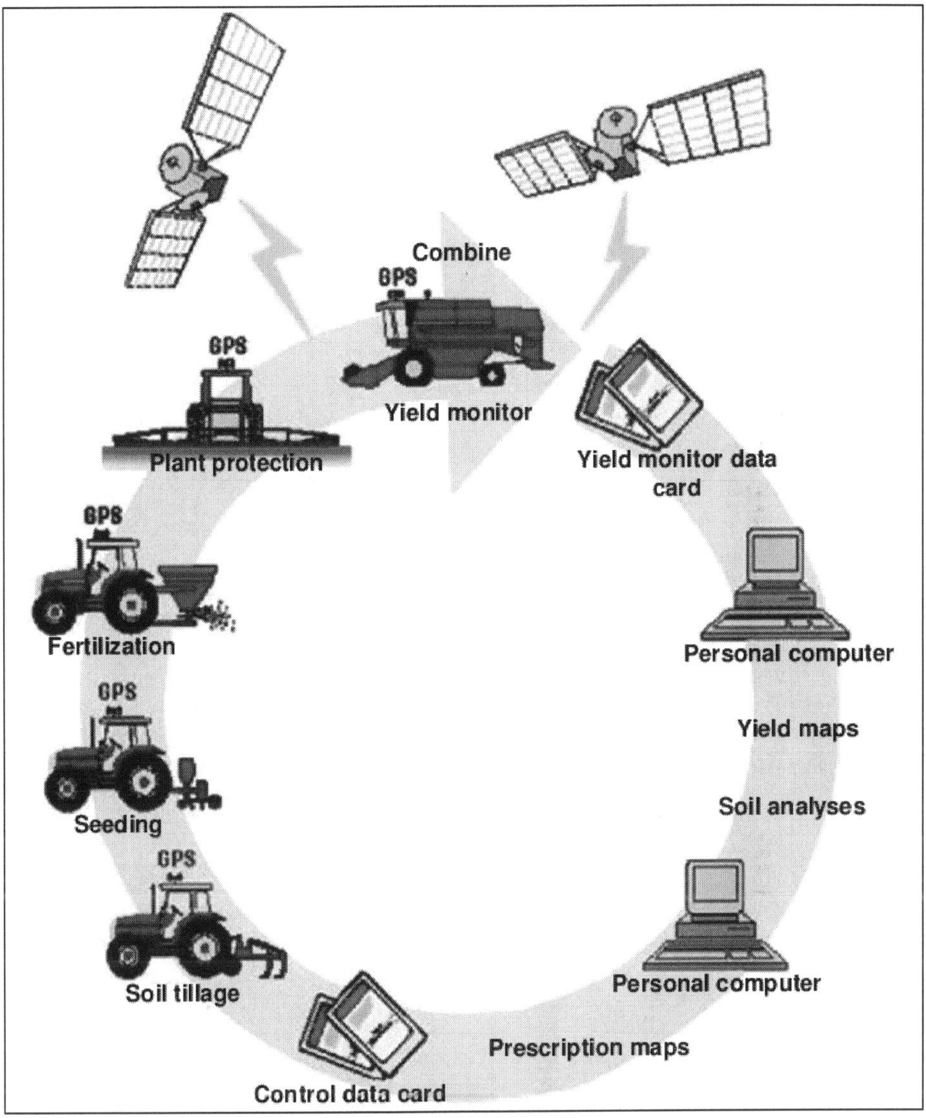

Figure 1.4. Diagrammatic Representation of Precision Farming Cycle.

A. Global Positioning System (GPS)

GPS provides continuous position information in real time, while in motion. Having precise location information at any time allows soil and crop measurements to be mapped. GPS receivers, either carried to the field or mounted on implements allow users to return to specific locations to sample or treat those areas. GPS satellites broadcast signals that allow GPS receivers to calculate their position. GPS allows farmers to locate the exact position of field boundaries and obstructions. GPS allows farmer to reliably identify field location so that inputs (seed, fertilizer, pesticide, herbicides) can be applied to an individual field based on perfomance criteria and previous input applications. GPS with combine yield monitor used to collect yield data across the land, GPS with variable rate planting system used to variable rate planting, GPS with variable rate controller used to variable rate time and fertilizer application and GPS with mapping software used to field mapping and insurance purposes.

B. Differential Global Positioning System (DGPS)

DGPS is a technique to improve GPS accuracy that uses pseudo range errors measured at a known location to improve the measurements made by other GPS receivers within the same general geographic area. It is operating in the 285 to 325 KHZ frequency range. DGPS is an enhancement to the GPS which provides improved location accuracy in the range of 15 meter nominal to about 1-3 cm in case of the best implementations. DGPS uses a network of fixed ground based reference stations to broadcast the differences between the position indicated by the GPS satellite system of known fixed position. **Pseudo range** is a actual distance between a satellite and navigation satellite receiver.

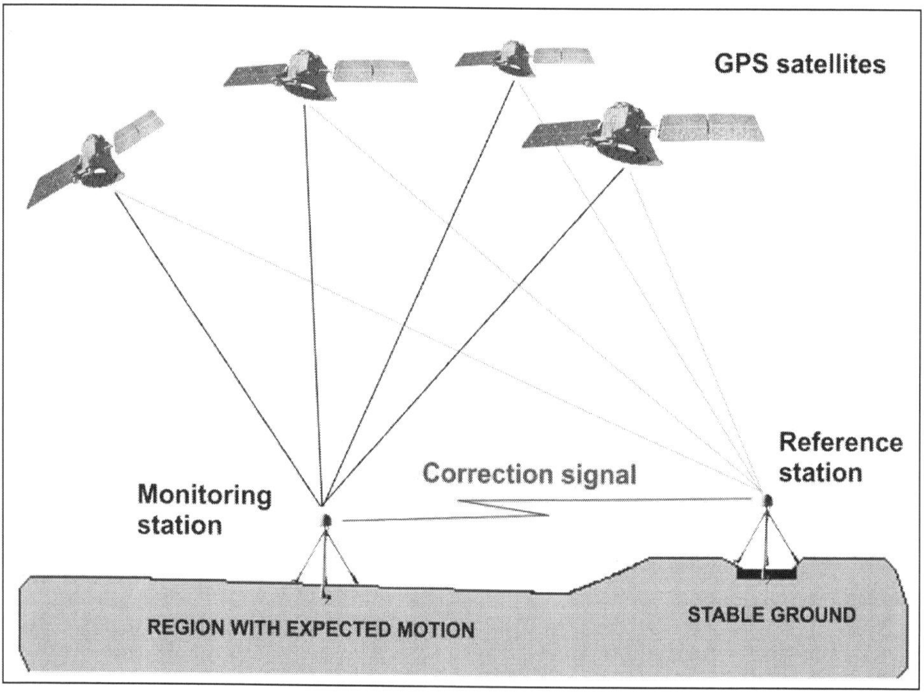

Figure 1.5. Operationals Principles of differential Global Positioning System.

C. Geographic Information Systems (GIS)

These are computer hardware and software that use feature attributes and location data to produce maps. GIS is store layers of information, such as yields, soil survey maps, remotely sensed data, crop scouting reports and soil nutrient levels. A farming GIS database can provide information on topography, soil types, surface drainage, sub-surface drainage, soil testing, irrigation, chemical application rates and crop yield. This information is used to understand the relationships between the various elements affecting a crop on a specific site. GIS can be used to evaluate present alternative management by combining and manipulating data layers to produce an analysis of management scenarios. Integration of the land and water resources and identification of the ecological problem at the micro level will help in identifying the location specific solutions through the effective use of remote sensing based resources information combined with other socio-economic data using GIS. It is useful in generation of action plan map giving site specific recommendations development and management of agriculture, ground water recharge, fuel, fodder, soil conservation, reclamation and afforestation.

D. Remote Sensing (RS)

It is the collection of data from a distance. Remote sensed data provide a tool for evaluating crop health, plant stress related to moisture, nutrients, compaction, crop diseases and other plant health concerns are often easily detected in overhead images. Remote sensing can reveal in-season variability that affects crop yield and can be timely enough to make management decisions that improve profitability.

E. Combine Harvesters with Yield Monitors

Yield monitors continuously measure and record the flow of grain in the clean-grain elevator of a combine. When linked with a GPS receiver, yield monitors can provide data necessary for yield maps.

F. Sensor Technologies

Sensors are physical devices, which respond to physical stimulus (heat, light, motion, pressure and sound) and transmit the impulse generated. Various technologies such as electromagnetic, conductivity, photo electricity and ultra sound are used to measure humidity, vegetation, temperature, texture, structure, physical character, nutrient level, vapors and air. Computer record the sensor impulse, a GPS measures the position and a GIS produce map and analyses the sensor data. Sensors enables the collection of immense quantities of data without laboratory analysis.

G. Variable Rate Technologies (VRT)

It allows fertilizer, chemical, irrigation water and other farm inputs to be applied at different rates across a field without manually changing rate setting on equipment. It is used for the application of various agricultural inputs (seeds, fertilizer, herbicides, pesticides) at a variable rate in the right place at the right time in order to respond adequately to the within field variability of environment factors like soil, pests and crop parameters. There are three main components of VRT; 1) Control computer for co-ordinates field operation with the help of database in its memory; 2) Locator receives the current location of equipment and issues command to the actuator, 3) Actuator for input application.

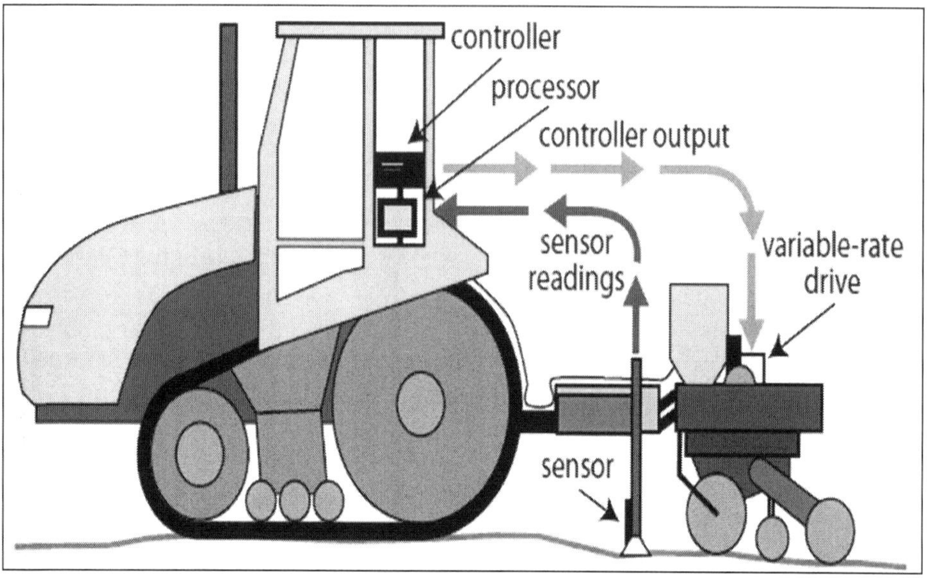

Figure 1.6. On the Go Sensor (Variable rate technology).

H. Variable Rate Applicator (VRA)

It is the process of applying different rates of crop inputs according to the variability within an agricultural field. Variable rate fertilizer application is a technology that regulates the fertilizer application rate based on site-specific needs within a field. The application map is loaded into a computer mounted on a variable-rate applicator. The computer uses the application map and a GPS receiver to direct a product-delivery controller that changes the amount or kind of product, according to the application map.

There are two types of VRA

1) **Map based:** A map based VRA adjusts the application rate based on an electronic map, also called a prescription map. Using the field position from a GPS receiver and a prescription map of desired rate, the concentration of input is changed as the applicator moves through the field.
2) **Sensor based:** Sensor-based VRA requires no map or positioning system. Sensors on the applicator measure soil properties or crop characteristics "on the go". Based on this continuous stream of information, a control system calculates the input needs of the soil or plants and transfers the information to a controller, which delivers the input to the location measured by the sensor. **VRA management zones** are the natural variability of a field and are used to manage the VRA of inputs across the

field. A **prescription map** is an electronic data file containing specific information about input rates to be applied in every zone of a field.

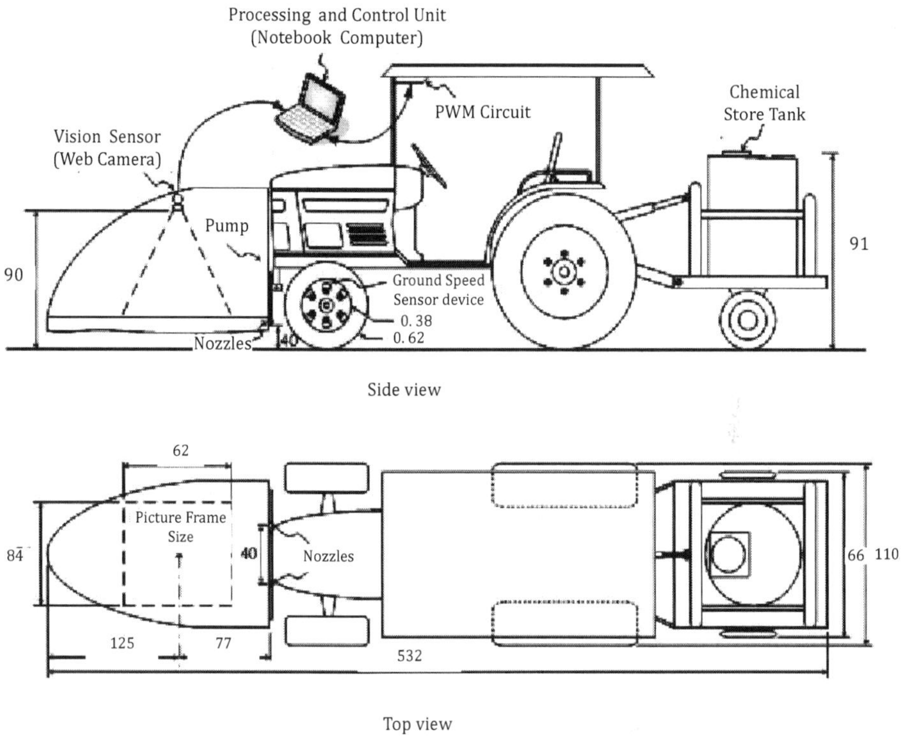

Figure 1.7. Schematic of a Variable Rate Applicator System.

I. Rate Controllers

These are devices designed to control the delivery rate of chemical inputs (fertilizer/pesticides). These rate controllers monitors the speed of the tractor/sprayer travelling across the field as well as flow rate and pressure of the material, making delivery adjustment in real time to apply a target rate.

J. Grid Soil Sampling

Soil samples may be collected for more than one area of field. These sample are analyzed in the laboratory and interpretation of crop nutrient needs then the fertilizer application map is plotted using the entire set of soil sample. The application map is loaded into a computer mounted on a variable rate fertilizer spreader. The computer uses the application map and GPS receiver to direct a product delivery controller that changes the amount of fertilizer product according to the application map.

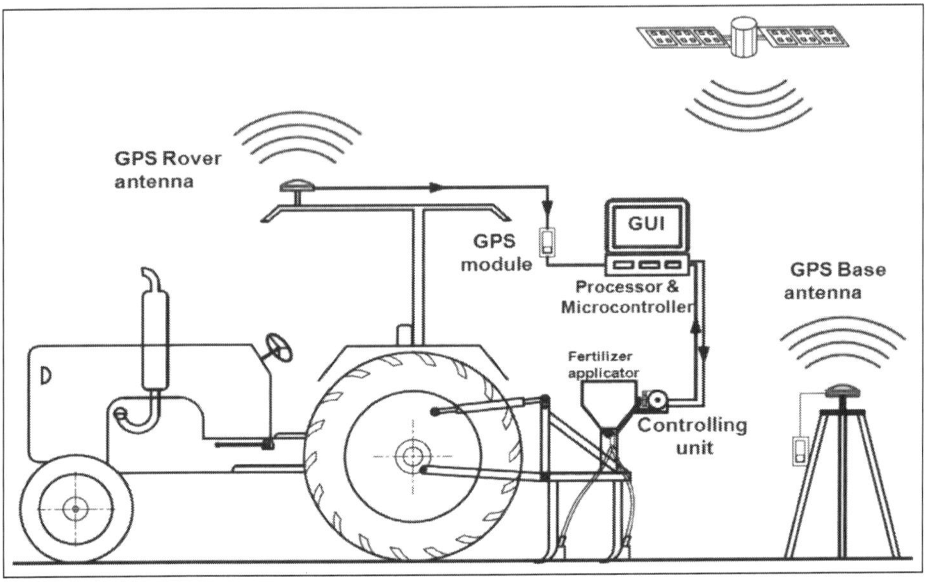

Figure 1.7. Diagram of Digital Map-based Variable Rate Fertilizer Application System.

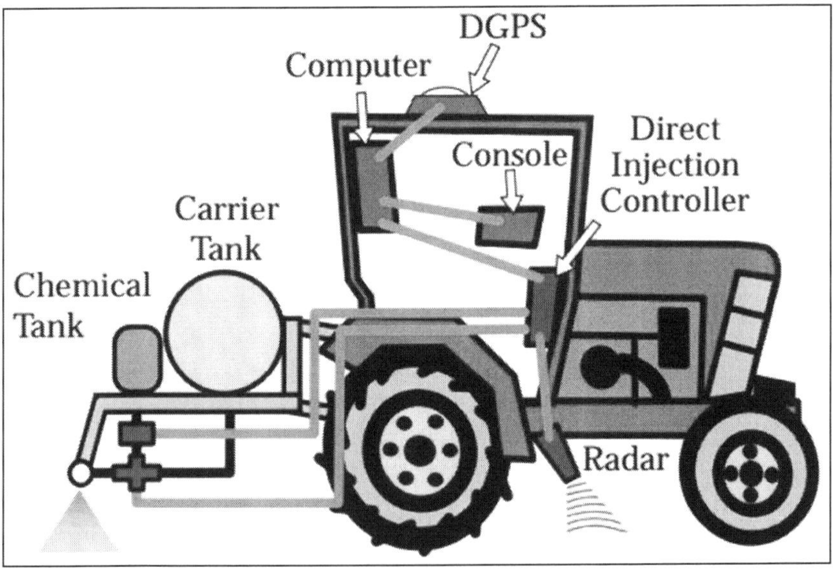

Figure 1.8. Components of the Variable Rate Technologies for Chemical Application.

K. Software

Appling precision agriculture technologies will frequently require the software to carry out diverse tasks. Software are used for; to filtering collected data, to generate variable rate application, to generate map and to provide advanced geo-statical features.

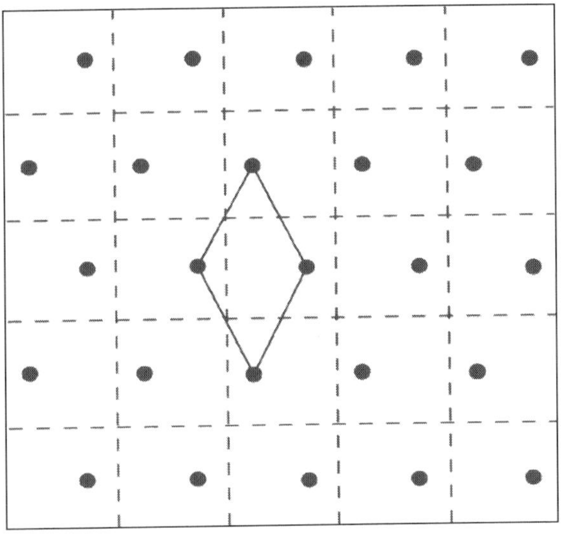

Figure 1.9. Systematic Grid - Diamond Sampling Pattern.

L. Yield Monitor

There are four main components of yield monitor; a) Sensors for measure the mass or volume of grain flow, separator speed, ground speed, grain yield is continuously recorded by measuring the force of the grain flow as it impacts a sensible plate in the clean grain elevator. b) Data storage device, c) user interface (display and key pad) and d) computer for controls the integration and interaction of all components.

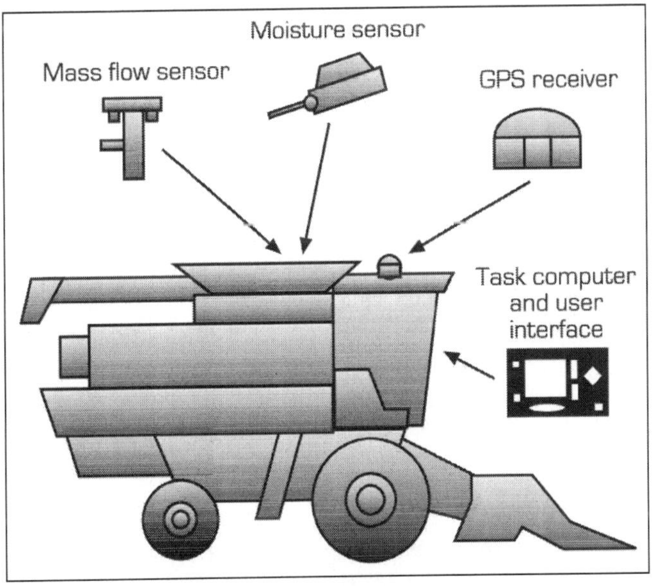

Figure 1.10. Components of a Yield Monitoring System.

1.6 Crop Management in Precision Farming

The crop management scheme of precision farming system comprises the following practical steps;

a) Determine management zone
b) Established yield goals
c) Soil sampling and soil testing
d) Data interpretation
e) Make decision regarding management of land preparation, varieties, fertilizers and other nutrients to achieve yield goals
f) Established map
g) Apply integrated pest management, integrated weed management and integrated disease management approach
h) Apply precision irrigation
i) Apply logging and automated record keeping
j) Monitor and established yield maps
k) Evaluate precision farming system response and identify strength and weakness for future improvement

1.7 Advantages and Disadvantages of Precision Farming

1.7.1 Advantages of Precision Farming

a) It will enhance agricultural productivity and prevent soil degradation in cultivable land resulting in sustained agricultural development
b) It will reduce excessive chemical usage in crop production
c) Water resources will be utilized efficiently under the precision farming
d) GPS allows agricultural fields to be surveyed therefore, the yield and soil characteristics can also be mapped
e) Dissemination of information about agricultural practices to improve quality, quantity and reduced cost of production in agricultural crops
f) It will minimize the risk to the environment particularly with respect to the nitrate leaching and groundwater contamination by means of the optimization of agro-chemical products
g) Non-uniform fields can be sub-divided into smaller plots based on their unique requirements. It provides opportunities for better resource management and hence reduce wastage of resources
h) Farmers can forecast and mitigate problems like water stress, nutrient deficiency, pests and diseases
i) Increases opportunities for skilled employment in the agricultural sectors

j) Provides new tools for evaluating multifunctional aspects including non-market functions
k) Essential for monitoring of greenhouse conditions in agricultural field

1.7.2 Disadvantages of Precision Farming

a) High capital costs may discourage farmers to not adopt this method of farming
b) Lack of technical expertise knowledge and technology
c) Not applicable or difficult or costly for small land holdings
d) Heterogeneity of cropping systems and market imperfections
e) Precision agriculture techniques are still under development and requires expert advice before actual implementation
f) It may take several years before the actual collection of sufficient data to fully implement the system
g) It is an extremely difficult task particularly the collection and analysis of data

1.8 Problems in Indian Agriculture

a) The small per capita land holding and land fragmentation
b) The cost of obtaining site specific data is high and not to the reach of marginal farmers
c) Heterogeneity of cropping system
d) The farmers seems to be less aware about the soil and environmental pollution causing due to use of chemical in agriculture
e) Imperfections in infrastructure and institutional constraints
f) Lack of farmers inclination to take the risk
g) Lack of technical expertise knowledge and technology
h) Unavailability of local technical expertise
i) Social and economic condition of farmers
j) Market imperfections
k) Demonstration of precision farming is not popularised
l) High initial investment
m) Enough success stories of precision farming are not available or published
n) Lack of tools and technology
o) Lack of technical centres for precision agriculture
p) Lack of specific software for precision agriculture

1.9 Approach Required to Promote Precision Farming

The following approach required to be adopted by the policy makers to promote precision farming at farm level;

a) Promote the precision farming technology for the specific progressive farmers who have sufficient risk bearing capacity as this technology may require capital investment.
b) Identification of niche areas for the promotion of crop specific organic farming.
c) Encourage the farmers to adopt water accounting protocols at farm level.
d) Promote use of micro level irrigation systems and water saving techniques.
e) Encourage study of spatial and temporal variability of the input parameters using primary data at field level.
f) Evolve a policy for efficient transfer of technology to the farmers.
g) Provide complete technical backup support to the farmers to develop pilots or models, which can be replicated on a large scale.
h) Pilot study should be conducted on farmer's field to show the results of precision agriculture implementation.
i) Policy support on procurement prices, in formulation of cooperative groups or self help groups.
j) Designation of export promotion zones with necessary infrastructure such as cold storage, processing and grading facilities.
k) Creation of multidisciplinary teams involving agricultural scientists in various fields, engineers, manufacturers and economists to study the overall scope of precision agriculture.

1.10 Difference Between Precision Farming and Traditional Farming

Sl.No.	*Precision Farming*	*Traditional Farming*
1	Farm field is broken into management zone	Whole field is treated as a homogenous area
2	Management decision are based on requirements of each zone	Management decision are based on field average
3	Precision farming tools are used to control zone	Inputs are applied uniformly across a field
4	Specific sowing of plant number and variety	Sowing with same plant number and variety
5	Nutrient management based on GPS and plant survey	Nutrient management based on average soil sample taking

Sl.No.	Precision Farming	Traditional Farming
6	Advanced technology used in crop production	Traditional methods followed in crop production
7	Machine operation adjusted to the arable site	Same machine operation practices
8	The field is regarded as heterogeneous arable site	The field is regarded as homogenous arable site

Chapter 2

Yield Monitoring and Mapping

CHAPTER OUTLINES

2.1 Definitions

Yield mapping is the process of continuously recording the grain flow through the combine harvester while at the same time integrating with actual harvest position in the field. The yield and position data is transferred to a computer where interpolation techniques are used to generate a yield map of the field. **or Yield mapping** refers to the process of collecting geo-referenced data on crop yield and characteristics, such as moisture content, while the crop is being harvested.

Yield monitoring is the estimation of crop yield well before the harvest. Aerial photography, radar, multispectral scanners and remote sensing data approaches are used for yield monitoring.

"A yield monitor is a great tool for collecting data and showing you what actually took place in a field. The interpretation of the data is what will lead you to change management practices."

— ***Grant Mangold***

2.1 Methods for Measuring Crop Yield

a) **Collect and weight method:** This is the oldest and simplest method. Harvested prop is weighed with the use of scales at a grain receiving facility or scale-equipped wagons. Farmers weigh and record the weight of each wagon of crop harvested from a field. The crop moisture content can be measured by sampling each weighed load. If the area from which each load was harvested is known then yield maps based on this type of collect and weigh data can be created. But in this case, a single average yield value is assigned to a larger field and this may not be appropriate for site specific farming purposes.

b) **Batch type yield monitors:** The system weighs the grain as it is loaded to the grain tank or as the grain tank is unloaded. The weight measurement may be displayed on an in-cab display. Both collect and weigh and batch type methods require calculation of area harvested to estimate the yield. However the yield values determined by this method are not site specific enough for the site-specific management.

c) **On the go yield monitors:** This type of systems continuously measures and records the yield as the crop is harvested. Most of these systems record each data point separately. It is also possible to measure and record the area harvested. If the data obtained with these systems are combined with a positioning system data, then site specific yield maps can be created for the purpose of site specific management. Most of these systems are also capable of measuring crop moisture content as the crop is harvested. The collect and weigh method can be used to calibrate this type of systems.

To determine instantaneous crop yield, the crop flow rate through the harvester's conveying system, the harvester's travel speed, the cutting width of the harvester's header and crop moisture content must be known. The area harvested per unit time is calculated using travel speed and cutting width then it is used with the volume or mass of the crop harvested to determine the yield.

A yield monitor consists of a flow sensor, a moisture content sensor, a ground speed sensor, a header position switch, a DGPS system, and a computer/display console. The schematic of the system components is given in the Figure 2.1.

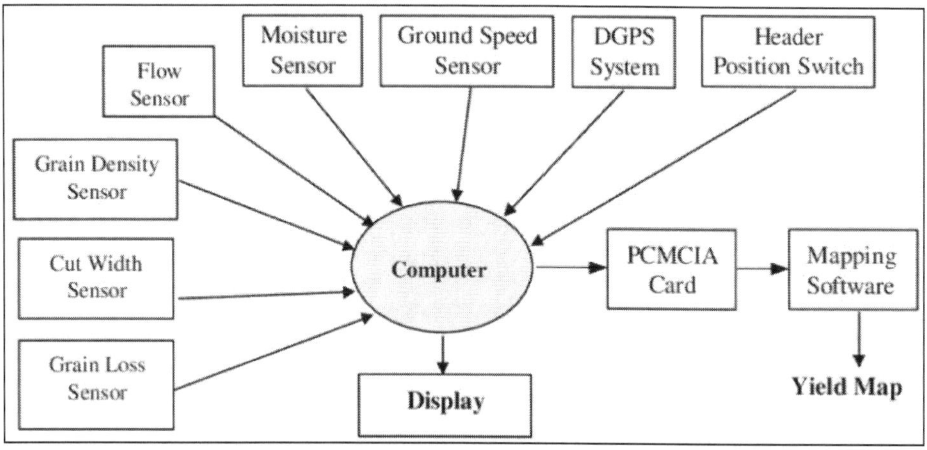

Figure 2.1. Components of a Yield Monitoring or Mapping System.

2.3 Components of Yield Monitor

2.3.1 Basic Yield Monitor Components

To determine instantaneous crop yield, three items must be known;

a) Flow rate through the harvester's separating system,

b) The harvester's travel speed and

c) The harvested width.

2.3.2 Flow Sensor

Commercially available yield sensors are based on **two different principles** for measuring the mass flow of grain;

i. **Volumetric:** A volume of grain is measured and then converted into mass by using the equation: **Density = Mass/Volume.** This system measures grain volume with the clean grain system. The **photo sensor monitor** uses a light source and a receiving sensor to detect the degree to which a combine's clean grain elevator is loaded. A photo sensor is a device used to detect light. The radiant energy is converted into an electrical signal. Measurements of light and dark periods by the photo sensor are used to estimate. When using the volumetric system, the variation of specific weight within the crop on measurement accuracy was highly significant. Volume measuring systems, accuracy is only guaranteed if regular measurements of the crop density are taken and manually programmed into the yield sensor's computer. The grain bulk density will change continuously with varying field conditions and volumetric systems require considerable effort in maintaining an accurate calibration.

ii. **Mass flow detection:** A mass flow sensor is a sensor that helps to provide the yield monitor with enough information to establish a grain yield measurement. The mass flow sensor works by using a load cell which is fixed to the top of a clean grain elevator. When the harvested grain is fed through the combine it eventually will hit up against this load cell, this is then transformed into an electrical signal and relayed to the yield monitor. The yield monitor uses this reading to determine how much grain is being taken into the combine at any point in time.

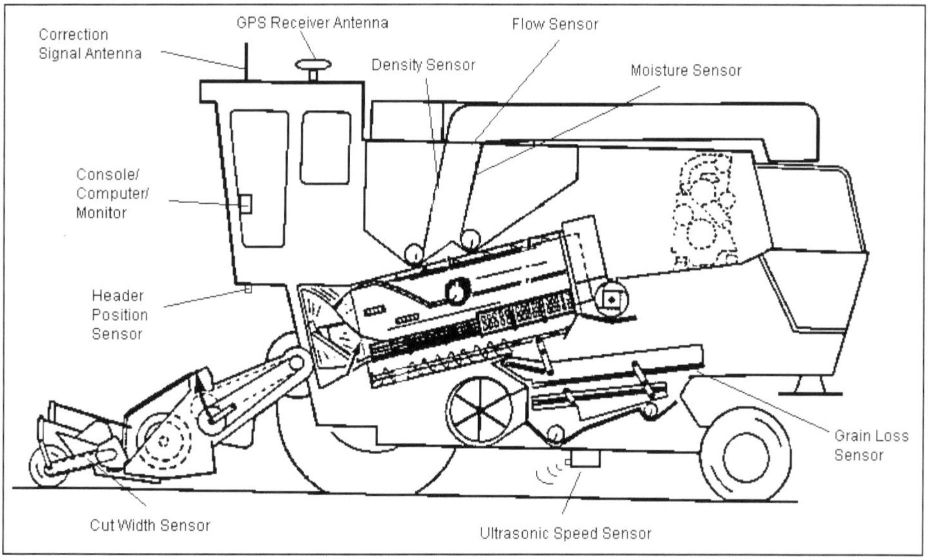

Figure 2.2. Combine Harvester Equipped with a Yield Monitoring System.

There are different methods used by flow sensors to measure crop flow on the harvesters to determine the yield these are;

a) **Impact force sensor**: Grain flow can be sensed using an impact plate. A force is applied by the grain impacting the plate when the grain strikes the spring-loaded plate. The force is proportional to the grain flow and is measured using a load cell. The load is converted into an electrical signal by the sensor.

b) **Plate displacement sensor:** It is also possible to measure the plate displacement by the use of a potentiometer. When grain strikes the spring-loaded plate, the impact plate is displaced. The displacement is proportional to the amount of grain that is impacting to the plate.

c) **Weight based sensor:** This type of flow sensors weighs the crop as it is conveyed in the harvester's conveying system. In essence, the variable that is measured is the change in the conveying system weight. As an example in combine harvesters, the grain is weighed as it passes through

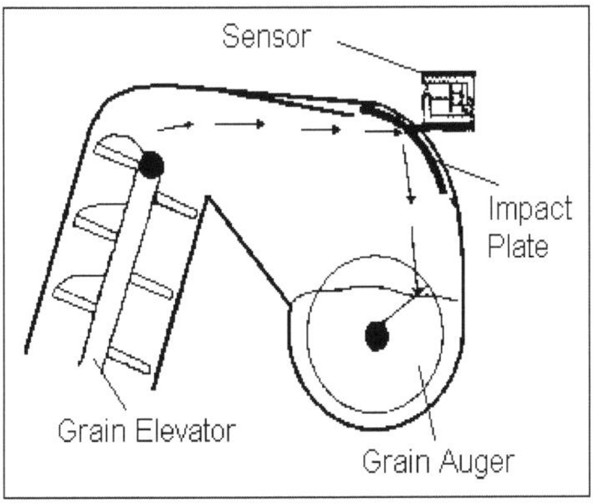

Figure 2.3. Impact Force Sensor.

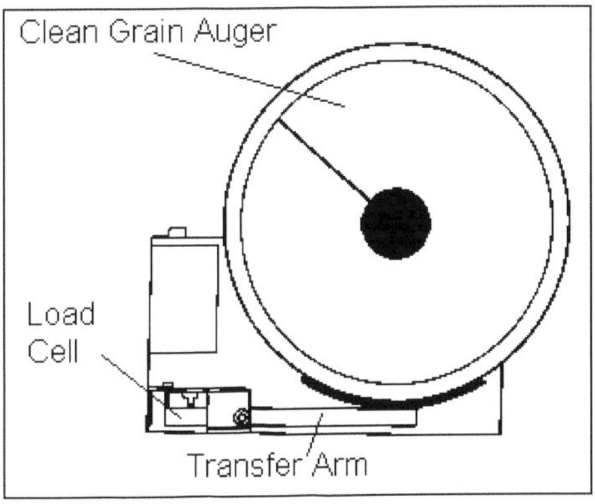

Figure 2.4. Weight-based Sensor.

the combine's clean grain auger. The change of the weight in the auger is sensed by a load cell. The change is proportional to the amount of grain passing through the auger. The radiometric sensor uses the relationship in determining the crop weight.

d) **Radiometric sensor:** This type of sensor consists of an isotropic or radioactive source and a detector. The source directs radiation toward a detector. Any obstruction between the source and the detector will reduce the intensity of radiation sensed by the detector. The reduction in intensity is dependent upon the amount of grain between the source and the detector.

e) **Photometric sensor:** This is a volumetric method. The system comprises a light source and a photo sensor. The light source emits lights and the photo sensor detects the light sent by the light source. Measurements of dark and light periods in the output signal of the photo sensor are used

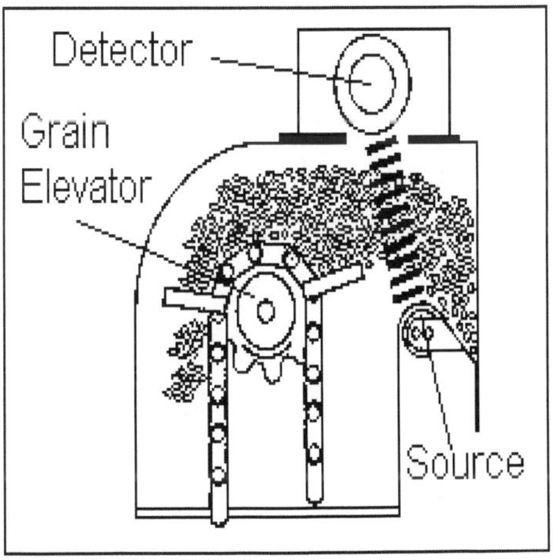

Figure 2.5. Radiometric Sensor.

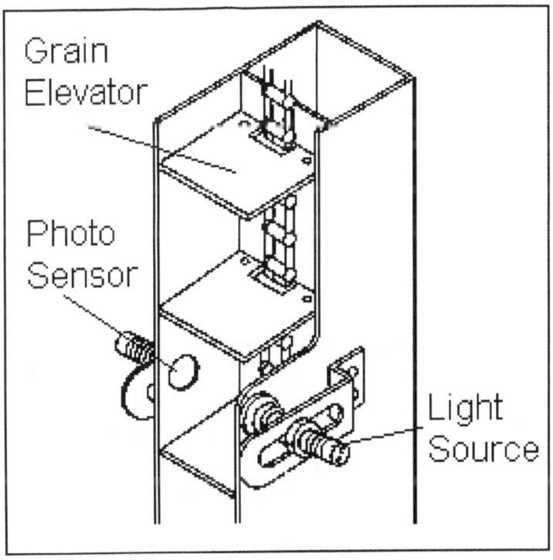

Figure 2.6. Photometric Sensor.

to estimate the crop volume flow rates. It is also necessary to measure the bulk density of the crop to calculate the yield since this method determines the volume of the crop.

f) **Paddle type sensor:** This is also a volumetric method and uses a photometry technique. The sensor consists of a paddle, a light source, a light detector called grain level detector and an electromechanical driving system to drive the paddle. When grain level goes high and prevents the light beam to reach the light detector, an electromechanical clutch engages a hydraulic motor that rotates the paddle wheel causing to discharge the grain until the grain level becomes lower than the light detector level. A reed switch determines the number of rotations leading the amount of grain since the volume of every division of the paddle is known, the volume of the crop is estimated from the number of rotations and the volume the paddle division.

g) **Force plate sensor**: This sensor applies a direction change on the grain with a curved plate or tube. The change in velocity is considered constant. The change in resultant force is related to the change in mass flow. More drain generates a greater force on the plate. A load cell can be used to measure the force and then it is related to the amount of grain mass.

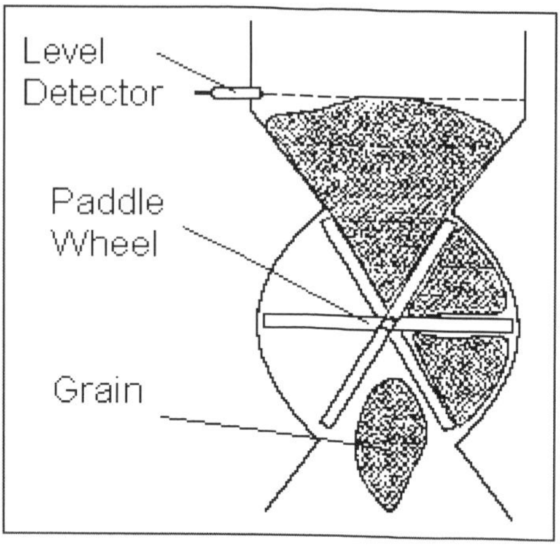

Figure 2.7. Paddle-type Sensor.

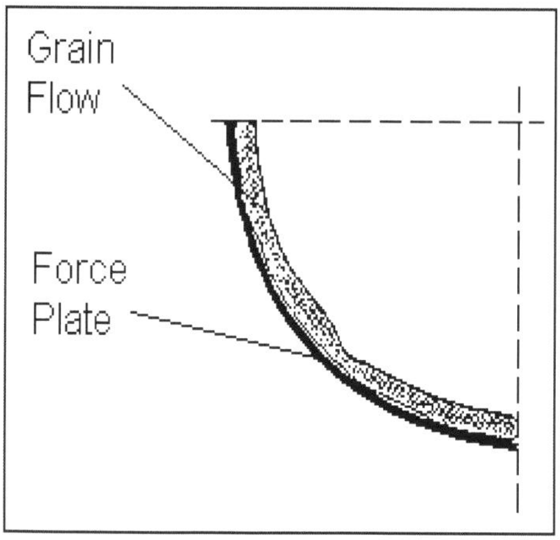

Figure 2.8. Force Plate Sensor.

2.3.3 Moisture Sensor

The moisture content of the crop being harvested affects the weight and the volume measurement of crop. When the moisture content is higher, the weight and the volume will also be higher. Therefore it is necessary to record moisture content at the time of harvest so that all yield data can be converted to a standard moisture value.

In general, a **capacitance type sensor** is used to measure the moisture content of the crop. Capacitors consist of two metal plates separated by a dielectric material. The sensor measures the dielectric properties of the crop that flows between the two metal plates. The higher the dielectric constant the higher the moisture content.

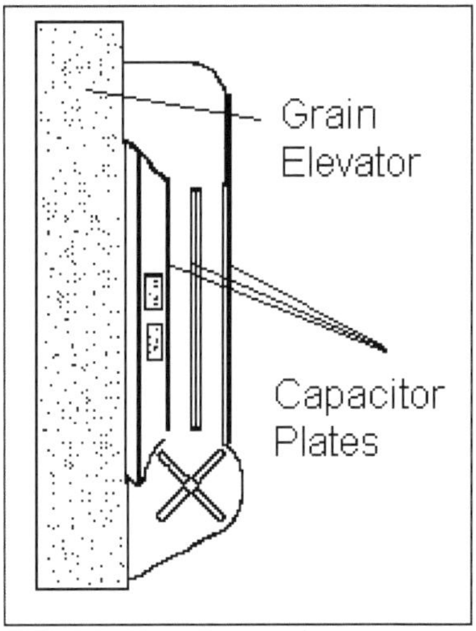

Figure 2.9. Capacitative Grain Moisture Sensor.

2.3.4 Ground Speed Sensor

It determines the distance the combine travels during a certain logging interval. It is measured using shaft speed sensor, radar and ultrasonic sensor and GPS based speed measurement.

a) **Magnetic shaft speed sensor:** The speed of the transmission output shaft speed of the combine is directly proportional to wheel speed. A magnetic sensor is used for the measurement. A magnet is attached to the shaft and a magnetic pick-up is mounted at a position having a distance of about 0.5 cm from the magnet. When the shaft rotates, the magnet causes a peak signal per one revolution. However the sensors may have errors due to slippage of combine drive wheels. Another error source is that the load on the harvester's tires increases as the tank of the harvester fills and this causes the tires to deflect and reduces the rolling radius of the drive wheels. These problems affect the accuracy of the ground speed measurement.

b) **Radar and ultrasonic sensors:** This type of sensors is more accurate than magnetic shaft speed sensor. Both types use guns that direct the signal toward the ground. Radar systems emit microwave signals while ultrasonic systems emit high frequency sound waves. The signals are reflected back to the sensor by the ground. There will be a frequency shift in the signal that returns to the speed sensor. The frequency shift is related to the ground speed and used for the speed measurement. In general, both systems are mounted on the harvester's frame close to the ground.

c) **DGPS based speed measurement:** A DGPS receiver is capable of calculating the speed of the harvester using instantaneous change in the receiver's position in the direction of its movement. The accuracy of the speed depends on the receiver's positional accuracy. A yield monitor obtains the speed data from a DGPS data sentence and interpret it to obtain the ground speed data.

2.3.5 Clean Grain Elevator Speed Sensor

It is used to improve accuracy of grain flow measurements.

2.3.6 Cutting Width Sensor

The theoretical cutting width of the combine harvester is known but in real harvest conditions, it is always less than that value.

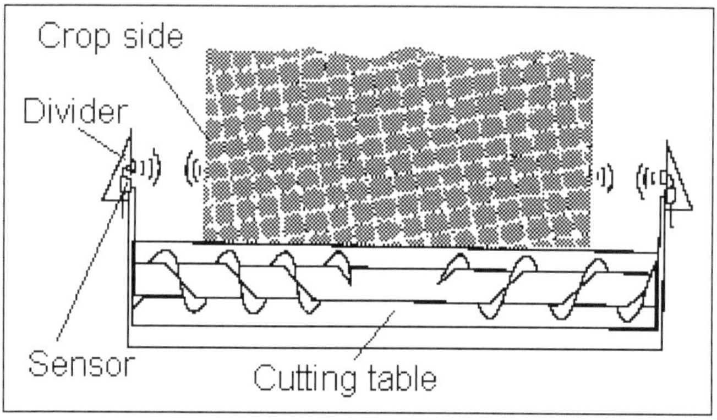

Figure 2.10. Ultrasonic Cutting Width Sensor.

For accurate yield monitoring, the current actual cut width must be known. The cutting width sensor measures the distance between the crop side and the cutting table side at both sides of the cutting table. The cut width should be less as the amount of the measured distance.

2.3.7 DGPS Unit

The DGPS unit determines the instantaneous position of the combine harvester as it operates in the field and provides this information in a computer compatible format. DGPS unit consists components:

i) **GPS receiver:** GPS receiver that gets signals from the satellites to determine the instantaneous position of the machine. GPS antenna receives satellite signal and

ii) **Differential correction signal receiver:** obtains the correction signal from a base station. Without a differential correction the position data would not be accurate enough. The purpose of using the differential correction receiver is to increase the accuracy of the position determination.

2.3.8 Grain Loss Sensor

For accurate yield monitoring, the amount of the grain that the harvester discharges to the field should also be accounted for by adding to the current yield value. The grain loss sensor measures the amount of grain that the harvester discharges. The sensor has an impact table mounted under the cleaning sieve. The grains impact to the plate. Every grain when hits the plate creates a peak signal and the counts of peaks in the signal and the appropriate calculation gives the amount of lost grain.

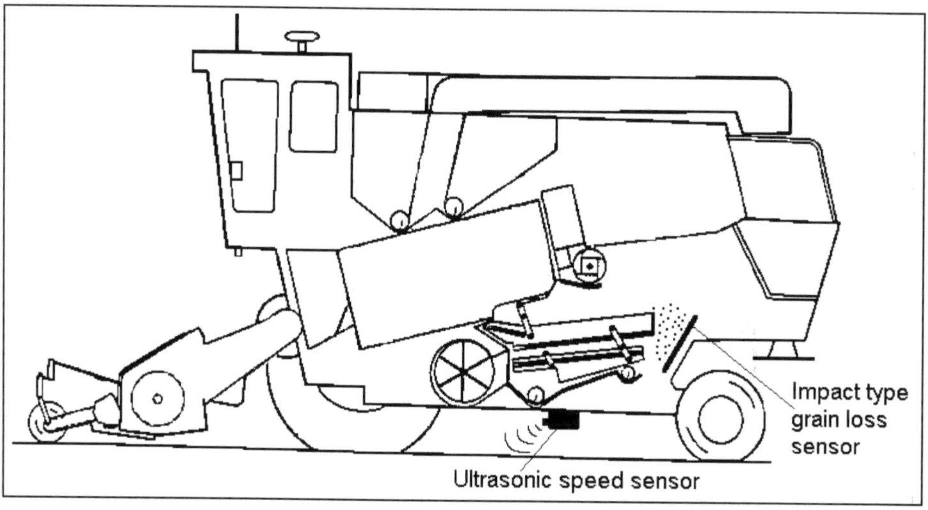

Figure 2.11. Ultrasonic Speed Sensor and Grain Loss Sensor.

2.3.9 Grain Density Sensor

The volumetric type flow sensors measures the volumetric flow of the grain. But the yield is expressed in mass unit and for this reason the volumetric yield should be converted into a mass unit by being multiplied by the density. The density can automatically be measured by a system.

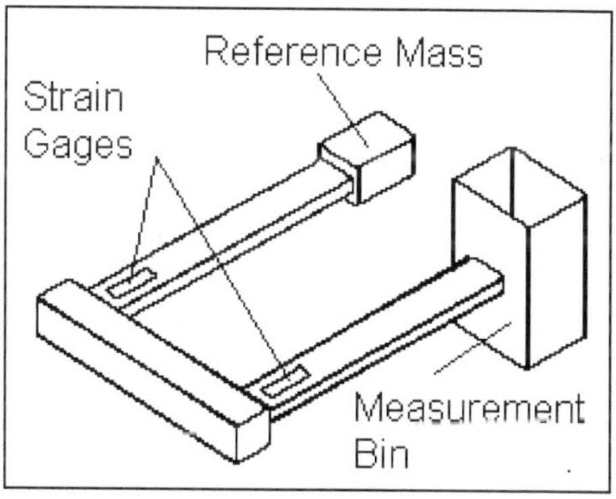

Figure 2.12. Grain Density Sensor.

The volume of the measurement bin is known and the mass of the bin is measured by using a strain gage. A reference mass is used to eliminate the effects of the vibrations and slopes.

2.3.10 Header Position Sensor

When there is no crop in the harvesting unit of the harvester when the header is in raised position, this sensor sends a signal to the computer stopping the area and yield calculation and when the header is lowered back, the yield calculation is resumed. This feature is used in turnings at row ends and other non-crop areas.

2.3.11 Harvester Position Switch

Controls the calculation of harvested acreage. When the sensor detects the header in the raised position, area counting is suspended. When lowered to a reasonable cutting height acre accumulation is resumed.

2.3.12 Display Console

The console connects to all of the sensors that supply information needed to calculate harvested yield. In addition to sensor inputs, the display console receives inputs from the combine operator. The operator can input, width of harvest, field or load information to permit tagging or referencing of the yield and moisture data.

2.4 Yield Map

Yield mapping is the first step to determine the precise location information of the highest and lowest yield areas of the field and to analyze the factors causing yield variation. Yield maps are based on sets of instantaneous yield data points collects with the yield sensor compotes and a DGPS receiver. Once the data is collected users need special computer software to generated yield maps. Most collected data is stored in a binary format. Binary files store digital data very efficiently, but this data must be converted to standard text files in order for most yield mapping software to process the data. The yield map provides information to the farmer detailing which are the better and poorer yielding parts of the field. The yield map does not provide information as the cause of the yield variation. The farmer is required to use a combination of local knowledge and other sources of data to interpret the maps to make crop management decisions to improve the profitability of the field.

2.5 Types of Errors

a. **Grain specific weight:** This error is more prevalent with volume type yield monitors. Systems based on the principal of mass flow detection are not susceptible to this error.

b. **Moisture sensor:** The measurement of grain moisture content along side grain yield is an important variable to consider to allow the measured weight of grain or yield to be corrected for a constant grain moisture level. The grain yield is normally corrected to 14 per cent moisture content. This becomes increasingly important when comparisons are to be made between yield levels in different parts of the field, between different fields, and for the same field between different seasons.

c. **Measured area:** The measured area is calculated by multiplying the cutter bar width by the distance traveled by the combine. The width of the cutter bar and rolling circumference of the wheel is entered into the combine's computer by the operator. As these values are not monitored during harvest, they are assumed to be constant until manually adjusted. If they deviate from the programmed values, then errors are introduced into the raw yield data set when converting grain flow, measured at the yield sensor, into grain yield.

d. **Delay for yield:** This represents the time delay from the crop being cut (true position), and traveling through the combine to the point where the grain flow is measured by the yield sensor.

e. **Lead-time:** There is a period of time that is required for the combine to fill with grain once it has entered the crop. The lead time is, the time lapse that is required for the combine to reach a stage when is has filled with grain and is giving a true reading on the yield sensor.

f. **Lag-time:** This is time required by the combine to empty of grain once is has stopped cutting the crop at the cutter bar. The lag time represents the time that true yield data can be logged once the combine's cutter bar has left the standing crop.

g. **Delayed positioning:** This is the offset of the GPS antenna. For practical reasons it is not possible to mount the GPS antenna directly over the cutter bar and it is normally mounted on the cab roof of the combine. The location of the GPS antenna on the cab roof introduces a "positioning offset".

h. **Turning on headlines without cutting any crop:** errors are recorded in raw data when the combine is turning on the headland with no grain being cut. In this instance a zero yield will be recorded in the data string, which will influence the processed yield map.

2.6 How to Reduce Errors

a. Re-calibrate the yield sensor, especially in the case of volumetric yield measurement

b. Check regularly for and remove any crop residue built up on the yield sensor

c. Always keep the cutter bar width full with crop when harvesting in straight lines

d. Reduce the number of narrow finishes by traversing the combine backwards and forwards from one side of the field

e. Always lift the cutter bar above the pre-determined height at the end of each harvest run to activate the automatic cut-off switch

f. Conduct repeated experiments to derive time delays for delay for yield, lead and lag to suit individual driving techniques and crop conditions

g. Reduce delay for yield errors by avoiding sudden changes in forward speed

h. Maintain a constant lead-time by lowering the cutter bar at the same point when entering the crop after a headland turn

i. Maintain a constant lag time by raising the cutter bar at the same point

j. Avoid mid-run stop/starts.

There are two types of yield monitors currently available;

a. Weight of grain when it is discharged into the grain tank. *e.g.* AgLeader, Micro-Trak and John Deere systems.

b. Senses the volume of grain as it is delivered to the grain tank. *e.g.* Satloc and Farmscan units. The Satloc unit measures the volume of grain as it passes up the clean grain elevator and Farmscan unit uses a turbine to calculate the volume of grain as it flows up the 'bubble up' auger.

2.7 Determination of Harvesting Yield

$$\text{Yield (Bushel/acre)} = K \times \frac{\text{Flow} \times \text{Time}}{\text{Width} \times \text{Length}}$$

where,

k = Unit conversion coefficient (Corn- 112011, Sorghum-104544)

Flow = Grain flow (pounds/seconds)

Time = Logging interval (Seconds)

Width = Swath width (inches)

Length = Distance traveled during the logging interval (inches)

The yield obtained can then be compensated for moisture variation using the following formula

$$\text{Yield compensated} = \text{Yield} \times \frac{100 - \text{Moisture}}{100 - \text{Moisture reference}}$$

Standard reference moisture value are 15.5 per cent for corn, 13 per cent soybean and sorghum and 12 per cent for wheat.

2.8 Benefits of Yield Monitoring

a. It give the user an accurate assessment of how yields vary within a field

b. Monitoring crop production and evaluating the economic benefits of management practices

c. Yield maps created with on-the-go yield monitoring systems

d. Provide useful information and enhance on-farm research

e. Yield data can be accumulated for a specific load or field, thereby facilitating the comparison of hybrids, varieties, or treatments within test plots

f. Get instantaneous readout in the field of accumulated grain weight, harvested area and average yield

g. Producer can conduct on-farm variety trials or weed control evaluations without the need of a weigh wagon

h. On-farm comparisons help producers fine-tune crop production practices to their soils

Chpater 3

Soil Mapping

CHAPTER OUTLINES

3.1 Definition

Soil mapping is the process of characterizing natural bodies of soils, classifying and grouping the defined soils into map units and capturing soil property information for interpreting and represent soil spatial distribution on a map.

3.2 Soil Survey

It is a study and mapping of soils in their natural environment **or** It is the systematic examination, description, classification and mapping of soils of an area. **or** It is the process of classifying soil types and other soil properties in a given area and geo-encoding such information. **or** It is the study of the natural environment with special reference to the soil resources, providing information on the spatial distribution of major soil types, their properties, potentials and responses to management. Primary data for soil survey are acquired by field sampling and remote sensing.

3.2.1 Soil survey comprises a group of interlinked operations involving;

a) **Field work:** To study the important characteristics of soils and the associated external land features such as landform, natural vegetation and slop *etc.*

b) **Laboratory analysis:** To support and supplement the field observations.

c) **Correlation and classification:** The classification of soils into defined taxonomic units.

d) **Mapping of soil:** It is establishing and drawing soil boundaries of different kinds of soils on standard topographical base map.

e) **Interpretations of soil survey reports:** For making predictions about the potential of soils for alternative uses like agricultural crops, grasses, fruit and ascertaining their management requirements for sustained production.

f) **Transfer of agro-technology** from the research stations to farmers fields through soil

3.2.2 Methods of Soil Survey

It is for preparing soil survey reports and soil maps.

a) **Grid soil survey:** The traverse line are located on a grid pattern and the density of mapping is adjusted according to the area surveyed. So that the number of observations per square centimeters of the final map is independent of the scale.

b) **Free soil survey:** The survey or chooses observation points on changes in physiographic as observed by aerial photo or satellite image.

3.2.3 Types of Soil Survey and Scale of Soil Maps

a) **Exploratory surveys:** These are rapid road traverses made to provide information about unknown regions. Scales vary from 1: 2,000,000 to 5,000,000.

b) **Reconnaissance surveys (Low intensity detailed surveys):** The established scale is 1: 250,000, but maps of 1: 500,000 to 1: 120,000 are included. They are usually integrated surveys making much use of aerial photograph interpretations. Mapping unit is mainly landform classes.

c) **Semi detailed surveys (Medium intensity detailed surveys):** The range is 1: 100,000 to 1: 25000, typically 1: 50000. Aerial photo interpretation is combined with substantial amount of field study. Mapping units vary from soil-landform classes to soil associations and series.

d) **Detailed surveys (High intensity detailed surveys):** Scale range is 1: 25000 to 1: 10000 inclusive and are produced mainly by field studies. Mapping units are soil series and phases of series.

e) **Intensive surveys:** Scales are larger than 1: 10000, usually 1: 2500 to 10000, typically 1: 5000. Grid or regular traverse methods of field survey are used. Mapping units are soil series and phases.

3.3 Purposes of Soil Mapping

a) Synthesize the key properties of the soil in a single document
b) Understanding of different soils as regards to their properties, genesis and classification
c) Making predictions about the behaviors of different soils for agriculture, forestry
d) Transferring technology by correlating the characteristics of soils of known behavior
e) Predicting the adaptability of soil to various uses and productivity
f) Provide a graphic documents for soil units, agricultural crops and variety
g) Appropriate application of fertilizers, manures as per soil fertility maps

3.4 Soil Map

Soil map is a geographical representation showing diversity of soil types, soil properties in the area of interest. It is a major product of soil survey. It is mostly used in developmental planning compared with other natural resource surveys. It applies the principles of soil science and draws heavily from geomorphology, theories of soil formation, physical geography and analysis of vegetation and land use patterns.

3.4.1 Map Scale

It is the ratio of distance on a map over the corresponding distance on the ground. The map scale is represented as **1: M** where, M is scale denominator. most of the GIS data are collected the common scale such as 1: 50000 or 1: 24000. Soil maps are required scales varies from 1: 1 million to 1: 4000 to meet the requirements of planning at various level.

a) Small scale **1: 1 million** are needed for macro level planning at national level;
b) Intermediate scale **1: 250000** are provide information for planning at regional or state level used for determine the suitability and limitations for several agricultural uses and requires intensity of soil observations;
c) Intermediate scale **1: 50000** are used for planning resources conservation and optimum land use at district level;
d) Large scale **1: 8000 or 1: 4000** these maps are generated through high intensity of field observation at village level.

3.4.2 Soil Mapping Units

a) **Soil series:** It represents a group of soils having similar horizons but they may differ in the surface texture.

b) **Soil types:** It is a sub-division of series based on the texture of the surface soil.

c) **Soil phase:** It is a sub-division of soil types in respect of soil slope, degree of erosion, stoniness, rockiness and salinity.

3.4.3 Methods of Interpretation of Soil Maps

a) **Visual image interpretation:** Visual interpretation is based on shape, size, tone, shadow, texture, pattern, site and association. Soil are surveyed and mapped using comprising aerial photograph, field survey and cartography. It is used in interpretation of elements like land types, vegetation, land use slope and relief. Soil mapping needs identification of these elements. It is used in impetration of not only aerial photographs but also multispectral satellite data. It is simple and inexpensive method of soil mapping.

b) **Computer Aided Approach:** It provide an efficient and accurate means of organizing, compiling, analyzing and displaying complex interrelated data. To analyse or numerical analysis of remote sensing data faster and extract information from the large quantities of data computer has been developed.

3.4.4 Advantages/Uses Soil Maps

a) Representation of diversity of soil types and soil properties in single documents

b) Used in land evaluation

c) Used spatial planning

d) Used environmental protection

e) Provide information on characteristics and condition of the land

f) Determine the suitability and limitation for agricultural uses

g) Determine planning resources conservation

Chapter 4

Geoinformatics

CHAPTER OUTLINES

4.1 Definitions

Geoinformatics is the science and technology of gathering, storing, analyzing, processing, interpreting, modeling and distributing geo-information. **or Geoinformatics** is the branch of sciences which uses the infrastructure of science to address the problems of geography, cartography, geosciences and related branches of science and engineering. **Agro-informatics** is the application in agriculture with innovative ideas, techniques and scientific knowledge to expand the horizons of the Computer Science. **Agro-geoinformatics** is the branch of geoinformatics about handling agro-geo-information.

4.2 Concepts of Geoinformatics

The term geoinformatics consists of two main words; **Geo** means earth surface and **Informatics** means facts about something. It is the multidisciplinary sciences that integrates the technology and principles of cartography, remote sensing, photogrammetry, surveying, global positioning system, geographical information systems and automated data capture system using high resolution geo-references spatial information from aerospace remote sensing platform. It provides tools that allows for the processing, manipulating and analysis of spatial data into information and use to make decisions about portion of the earth and environmental problems. It is modern technology that provide accurate means of measuring the extent and pattern of changes and other related information about environment.

For understanding the concepts of geoinformatics following principles of the compounds was important;

a) **Cartographic principles:** involves map, map design and production in analogue and digital computer environment,

b) **Remote sensing principles:** involves gathering of information of the environment without physical contact with object sensed by using electromagnetic energy radiation, interaction and detection principles in analogue or digital forms,

c) **Photogrammetric principles:** involves the art and scientific process of obtaining reliable information about the physical environment by interpreting remotely senses aerospace data in analogue and digital form,

d) **Surveying principles:** determines the precise position and dimension of point on the earth surface,

e) **Global positioning system principles:** determination of position of points presentation of the results in map and tables,

f) **Geographical information systems principles:** involve data gathering, data processing database management, modeling in a digital environments,

g) **Automated data capture systems principles:** include multi-spectra remote sensing processes, GPS data, maps and scanning and computer input and output technologies.

4.3 Tools and Techniques of Geoinformatics

a) Geodesy

b) Remote sensing

c) Geographic information systems

d) Global positioning systems

e) Cartography

f) Photogrammetry

g) Surveying

4.3.1 Geodesy/Geodetics

It is the discipline that deals with the measurement and representation of the earth. Geodesy is defined as the science concerned with the study of shape and area of the earth.

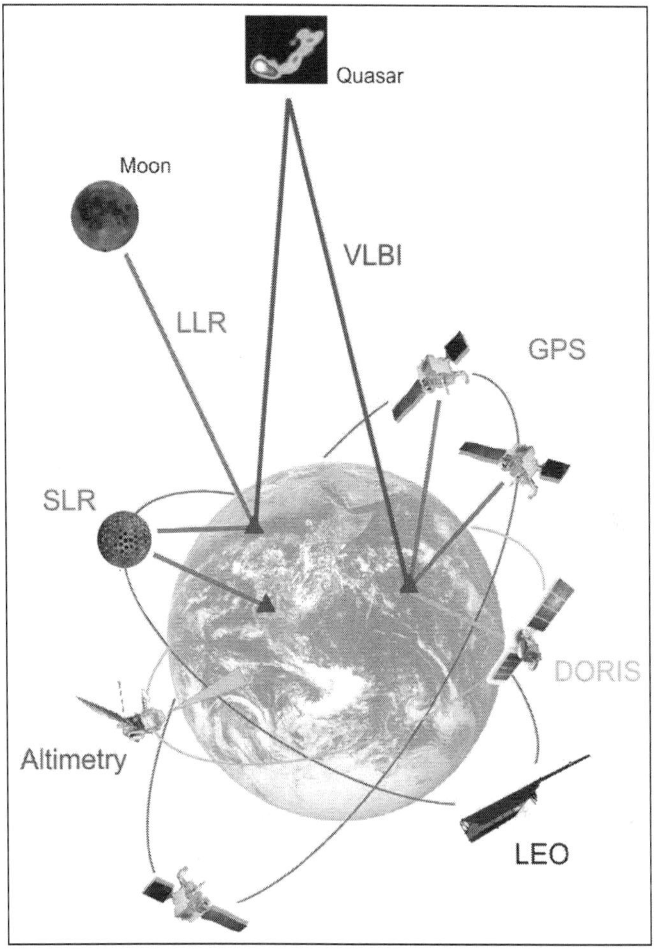

Figure 4.1. Different Methods of Satellite Geodesy.

Geodesy defines the shape and dimensions of the earth through its two branches;

i) **Gravimetry** deals with the determination of earth's gravity and its anomalies and the gravity determines the shape of the earth and

ii) **Positioning astronomy** determines the position of the points on the globe through the observation of stars and artificial satellites. The concept of geodesy helps in representing the earth in 2-dimensions. Satellite Geodesy is dealing with the observation and measurement of the earth from space. The space geodetic techniques are global positioning system (GPS), very long baseline interferometry (VLBI), satellite laser ranging (SLR), lunar laser ranging (LLR), doppler orbitography and radiopositioning integrated by satellite (DORIS), low earth orbiters (LEO) and altimetry.

4.3.2 Remote Sensing

It is the process of gathering of information of the environment without physical contact with object. Sun is the principal source of energy. When the energy (in the form of electromagnetic radiation) reaches the earth's atmosphere, it undergoes the process of reflection, absorption and transmission. Earth's surface consists of different natural and man-made features which reflect, absorb, store and emit earth's radiation at different wavelengths in different percentages, depending upon their physical and chemical properties.

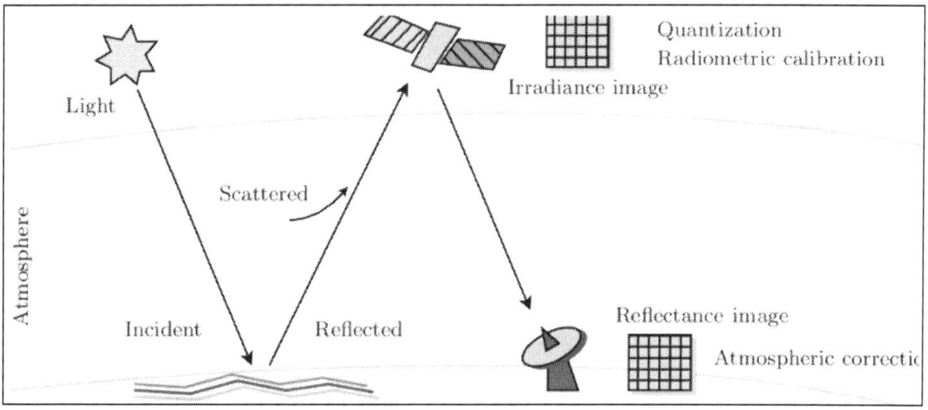

Figure 4.2. A Schematic Depiction of the Remote Sensing Process.

Remote sensing technology applicable to the observation of following parameters;

a) **Applicable to crop survey**- crop identification, crop area, crop vigour, crop density, stages of growth, crop growth rates, yield forecasting actual yield, soil fertility,. pest, diseases incidences, irrigation requirement, water quality and location of canals.

b) **Applicable to range survey** - forest type, forage capacity, seasonal change time, location of water. water quality, soil fertility-moisture, insects infestation and wild life inventory.

c) **Applicable to livestock survey-** cattle population, sheep- pig -poultry population, age-sex distribution animal behaviors, disease identifications and types of farm builds. Remote sensing imagery has many applications in mapping land-use and land cover, agriculture, soil mapping, forestry, city planning, archaeological investigations, military observation, and geo-morphological surveying, mineral exploration, among other uses.

4.3.3 Geographic Information System

It is a system for capturing, storing, manipulating, analyzing and displaying data which are spatially referenced to the earth. This system comprises hardware, software, database, procedures and expertise designed to support the storage, manipulation, analysis and display data of feature attributes and location data to produce maps. From these, information on vegetation, soil types, plant stress, crop infestation, information on agricultural land use planning are collected for the purpose of timely decision making. GIS has been used for emergency planning, logistics and transportation related analysis, crime analysis, business and service planning, government and public services, analysis of genome sequences on DNA, *etc.*

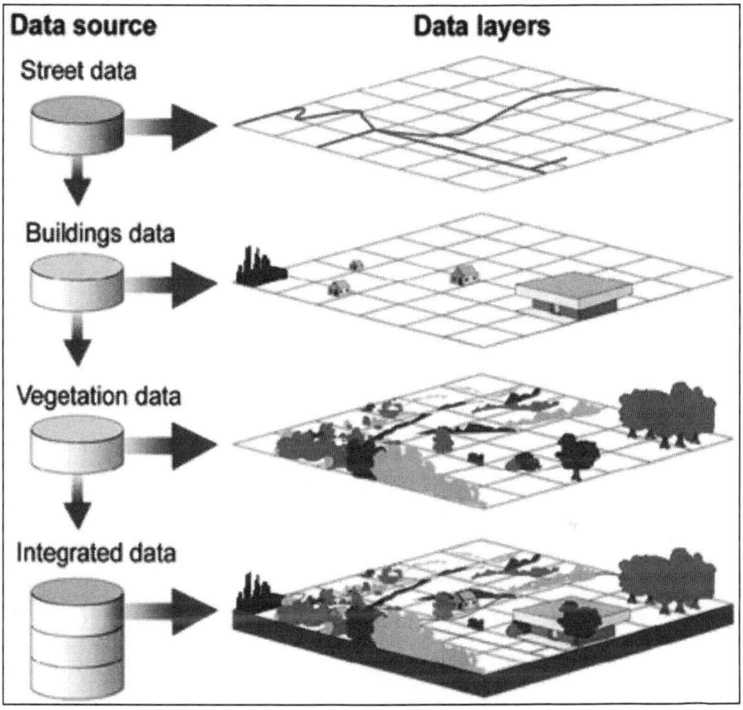

Figure 4.3. Different Layers in a GIS Map.

4.3.4 Global Positioning System

This system comprises ground control segments, space segment and user segment to provides geo-location. GPS allows the accurate location of soil samples within the field hence, the determination of physical, chemical and biological characteristics of the soil at different location and also fertility levels can be mapped across the field to serve as a basis for the application of farm inputs. GPS is also collected yield data at accurate location. It is needed for the production of yield maps and yield monitoring. Applications of GPS fall into following categories:

a) **Location:** determining a basic position;

b) **Navigation:** getting from one location to another;

c) **Tracking:** monitoring the movement of people, animals and goods
d) **Mapping:** creating maps of the world.

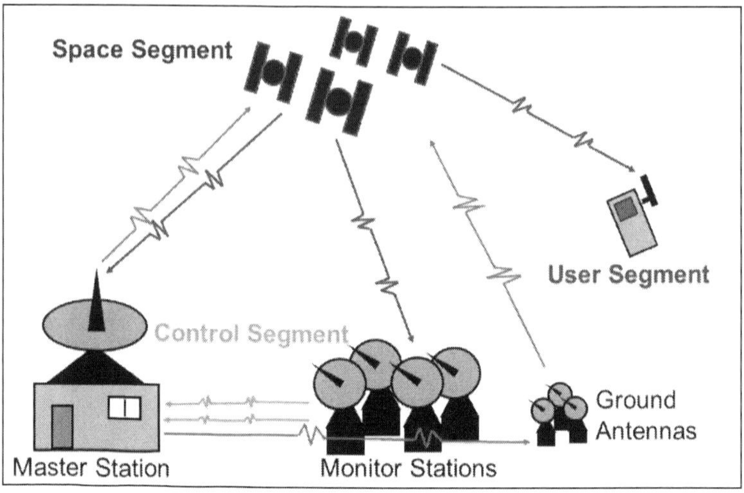

Figure 4.4. Segments of Global Positioning System.

4.3.5 Cartography

It is an art, sciences or technology of designing, constructing and producing maps. International Cartographic Association (founded on June 9, 1959, in Bern, Switzerland) defines Cartography as the discipline dealing with the conception, production, dissemination and study of maps. **Map** is a drawing of the whole or

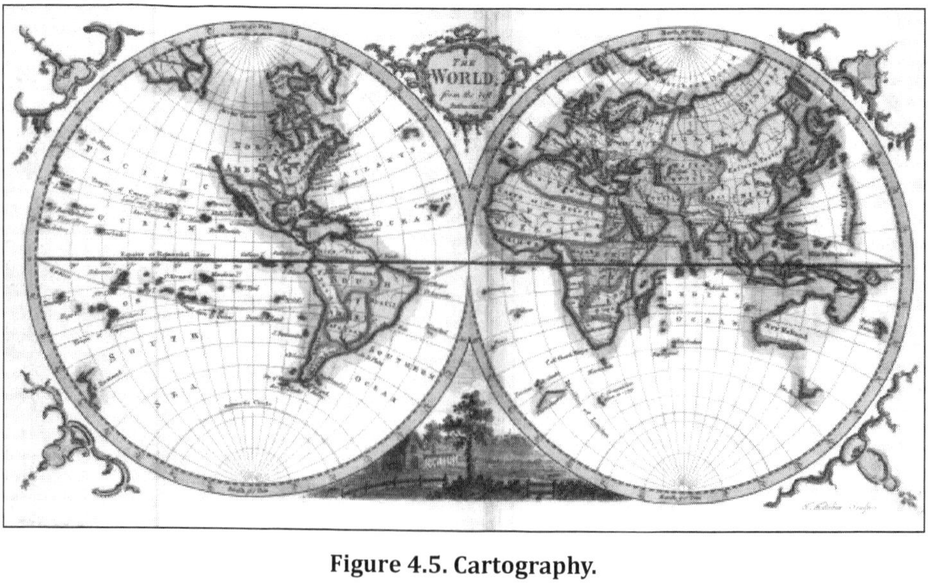

Figure 4.5. Cartography.

part of the surface of the Earth on a plane surface to a particular scale. Cartographic representation is the key in deciding the fate of any map generated out of various geoinformatics analysis. **Cartographic visualization** is the graphical presentation of geographic information, such as data, processes, relations or concepts, limits the extent of map details which could be incorporated for the given study area.

4.3.6 Photogrammetry

The American Society for Photogrammetry and Remote Sensing (ASPRS NGO Founded in 1934) has defined **photogrammetry** as the art, science and technology of obtaining reliable information about physical objects and the environment through processes of recording, measuring and interpreting photographic images and patterns of recorded radiant electromagnetic energy and other phenomena.

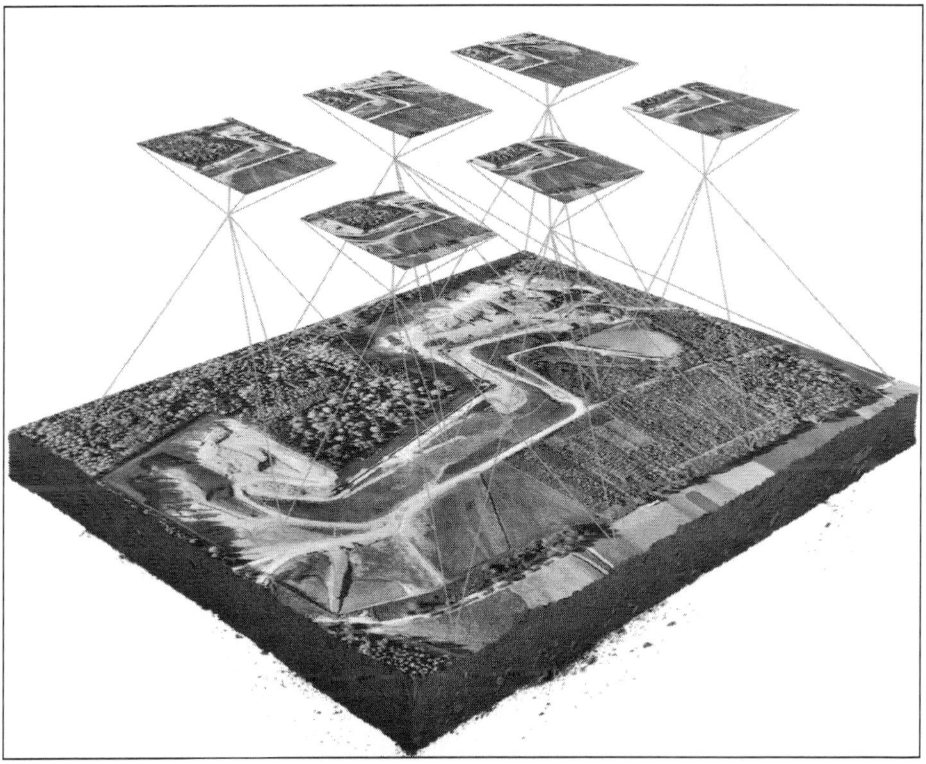

Figure 4.6. Photogrammetry.

Photogrammetry is useful in various fields including topographic mapping, architecture, engineering, manufacturing, quality control, police investigation and geology. Archeologists use photogrammetry to produce plans of large or complex sites. Meteorologists use it to determine the actual wind speed of a tornado in places where objective weather data cannot be obtained.

4.3.7 Surveying

Surveying is concerned with the measurement of locations of objects on the earth's surface, particularly property boundaries. Surveying provides high quality data on positions of land boundaries, buildings, *etc.* Land surveying is a technique and science of accurately measuring the distances and angles between different points, on the surface of earth. Aerial survey is a form of collection of geographical information using airborne vehicles. The collection of information can be made using different technologies such as aerial photography, radar, laser or from remote sensing imagery using other bands of the electromagnetic spectrum, such as infrared, gamma, or ultraviolet.

4.4 Geoinformatics for Precision Farming

Precision farming uses information based farm management for optimum profitability, sustainability and resource utilization. The required information for Precision farming is obtained with the help of technologies like GIS, GPS, RS, *etc.* **Geoinformatics** deals with integrating computer science and geosciences to solve complex scientific questions. The success of precision farming relies on the integration of these technologies into a single system that can be operated at farm level with sustainable effort.

a) **Remote sensing:** RS is the science of making inferences about material objects from measurements, made at distance, without coming into physical contact with the objects under study. The RS system consists of a sensor to collect the radiation and a platform an aircraft, balloon, rocket, satellite on which a sensor can be mounted. Remote sensing is used in precision farming for crop yield modelling, identification of pests and disease infestation, soil moisture estimation, irrigation monitoring, assessment of crop damage, *etc.*

b) **Global positioning system:** GPS provides continuous, real-time, 3D positioning and navigation worldwide in any weather condition. GPS-based applications in precision farming are being used in farm planning, field mapping, tractor guidance, variable rate applications (automated precise application of pesticides, fertilizers, *etc.* based on data that is collected by sensors, maps, and GPS) and yield mapping. GPS allows farmers to work during low visibility field conditions such as rain, dust, fog, and darkness. **Differential global positioning system:** Used for precise location of activities.

c) **Sensor technology**: Provide data on soil properties, crop condition, yield, fertilizer flow and weed detection.

d) **Geographic information system:** The GIS is a computerised data storage and retrieval system, which can be used to manage and analyse spatial data relating crop productivity and agronomic (study of crops and the soils) factors. It can integrate all types of information and interface with other decision support tools. GIS can display analysed information in maps that allow better understanding of interactions among yield, fertility, pests, weeds and other factors, and decision-making based on such spatial relationships.

e) **Variable rate applicator:** Used to operationalise precision farming at the farm level.

f) **Variable rate technologies:** Used to apply a desired rate of crop production materials at a specific time at specific location.

4.5 Advantages of Geoinformatics

a) It can acquire geospatial data in time and cost effective manner

b) It can provide synoptic view and information of an area in different parts of EMS

c) It helps in collecting data of an area without disturbing or destroying it

d) It assists in predicting events/outcomes based on multi-criteria analyses

e) It helps in visualizing the scenarios under a given situation and its consequences. e. g. scenarios greatly assist disaster managers to plan for future and save human lives

f) It helps to create a comprehensive solution to a problem in hand particularly assisting the decision and policy makers to arrive at an appropriate decision

g) It assists in planning strategies and useful in disastrous events for quick decision making

Chapter 5

Geographical Information Systems

CHAPTER OUTLINES

5.1 Definitions

GIS is a computer based system designed to capture, store, manipulate, analyze, manage and present all types of spatial or geographical data. **or GIS** is a powerful set of tools for collecting, storing, retrieving, transforming and displaying of spatial data form the real world (Burrough). **or GIS** is an information technology which stores, analyses, and display both spatial and non-spatial data (Parker). **or GIS** is a computer system capable of assembling, storing, manipulating, and displaying geographically referenced information. (USGS, 1997). **or GIS** is a set of computer based systems for managing geographic data and using those data to solve spatial problems (Lo and Yeung). **or GIS** is a computer system that allows the analysis and display of data with a spatial component (Phillips).

The GIS was developed in Canada during 1960 to analyze the Canada Land Inventory. The first known use of the term "geographic information system" was by **Roger Tomlinson** in the year 1968 in his paper "A Geographic information system for regional planning". **Roger Tomlinson** is also acknowledged as the **"Father of GIS"**.

5.2 Objectives of GIS

a) Maximize the efficiency of planning and decision making

b) Provide efficient means for data distribution and handling

c) Elimination of redundant data base minimize duplication

d) Integrate information from many sources

e) Complex analysis involving geographical referenced data to generate new information

5.3 GIS Subsystem and Functions

A GIS software has four main functional subsystems. These are: data input; data storage and retrieval; data manipulation and analysis; and data output and display subsystem.

a) Data input: A data input subsystem allows the user to capture, collect, and transform spatial and thematic data into digital form. The data inputs are usually derived from acombination of hard copy maps, aerial photographs, satellite images, remotely sensed image and GPS data.

b) Data storage and retrieval: The data storage and retrieval subsystem organizes the data, spatial and attribute, in a form which permits it to be quickly retrieved by the user for analysis and permits rapid and accurate updates to be made to the database. This component usually involves use of a database management system (DBMS) for maintaining attribute data. Spatial data is usually encoded and maintained in a proprietary file format. A collection of non redundant data which can be shared by different application systems is known **database**.

c) Data analysis and manipulation: The data manipulation and analysis subsystem allows the user to define and execute spatial and attribute procedures to generate derived information. This subsystem is commonly thought of as the heart of a GIS and usually distinguishes it from other database information systems and computer-aided drafting (CAD) system.

d) Data output (display or visualization): The data output subsystem allows the user to generate graphic displays, normally maps, and tabular reports representing derived information products.

Data capture, storage, query, analysis, display and output/presentation or visualization are the basic functions of GIS.

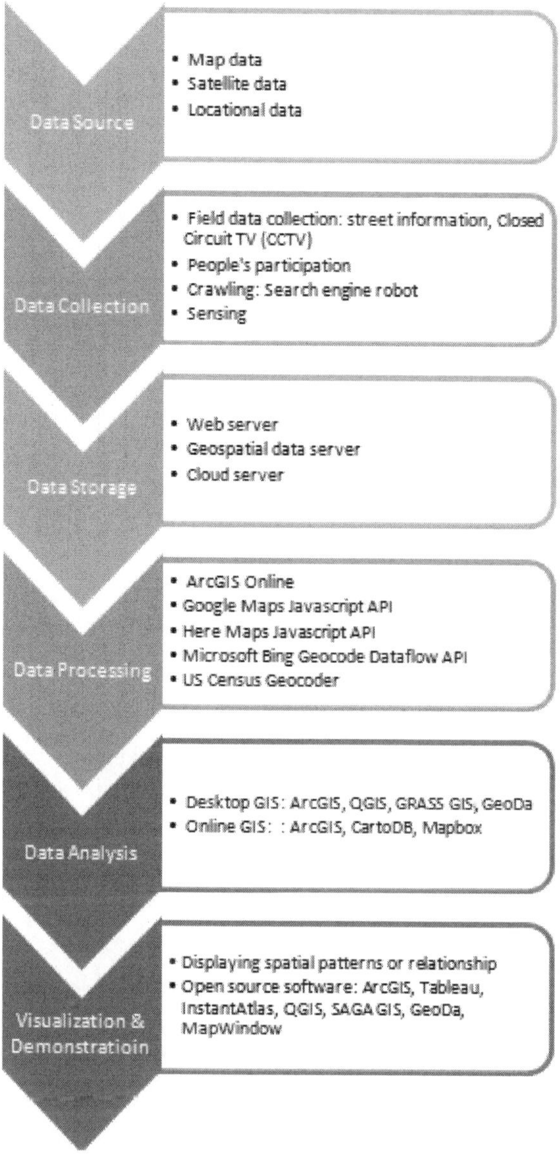

Figure 5.1. GIS Data Processing and it's Elements.

5.4 Components of GIS

A. Hardware

It consist of technical equipments *i.e* **computer** which has good processing capability to run the GIS software. These computer should have enough **hard disk** space to store large spatial data, tape drives and floppy drives. It consist of **input and output devices** like scanners, printers, plotters, digitizers, keyboard and monitor.

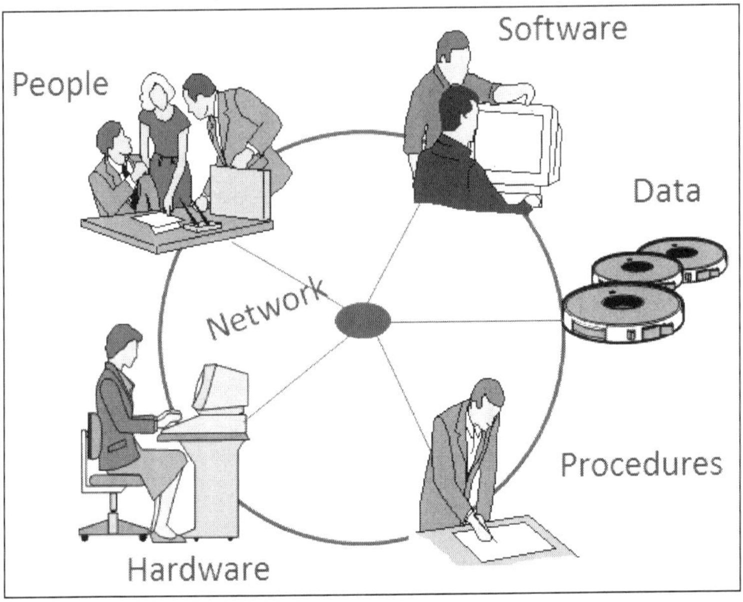

Figure 5.2. Components of GIS.

B. Software

GIS software provides the function and tools needed to store, analyze and display geographic information in the form of maps and reports, provides the a graphical user interface (GUI) for easy display. It is tools for input, visualizing, processing, editing, analyzing and queering geographical data. Data is accessed and managed through data base management system.

C. Data

Geographic data and related tabular data are the backbone of GIS. A GIS is integrate geographical data (spatial data), attribute data, remote sensing data, global database and related tabular data (non spatial data) by using DBMS (Data base management system).

D. Procedures

To guide its operations

E. Expertise/People

Technical persons who design and maintain the GIS work and use it. a. Viewers - The people who just use GIS for reference, b. General users - Use GIS for business, service and making decision such as planners and c. GIS specialist are the backbone of GIS and maintain, process and analyze geographic data.

5.5 Types of GIS Software

1. Arc GIS	2. QGIS	3. SAGAGIS
4. gVSIG	5. IDRISI	6. GRASS GIS
7. u Dig	8. MapInfo	9. Global Mapper
10. GeoMedia	11. JUMP GIS	12. Capaware
13. Geomatica	14. ENVI	15. ILWIS

1980 Rise of commercial GIS. Environmental system Research Institute (ESRI), CARIS- Computer Aided Resource Information Systems (CARIS). MapInfo Corporation and Earth Resource Data Analysis System (ERDAS) emerged as commercial vendors of GIS software. Arc GIS software is hold 80 per cent global market.

5.6 Types of GIS Data

a) **Numeric data:** It is statistical data which includes a geographical component that can be joined with vector files so the data can be queried and displayed as a layer on a map in a GIS.

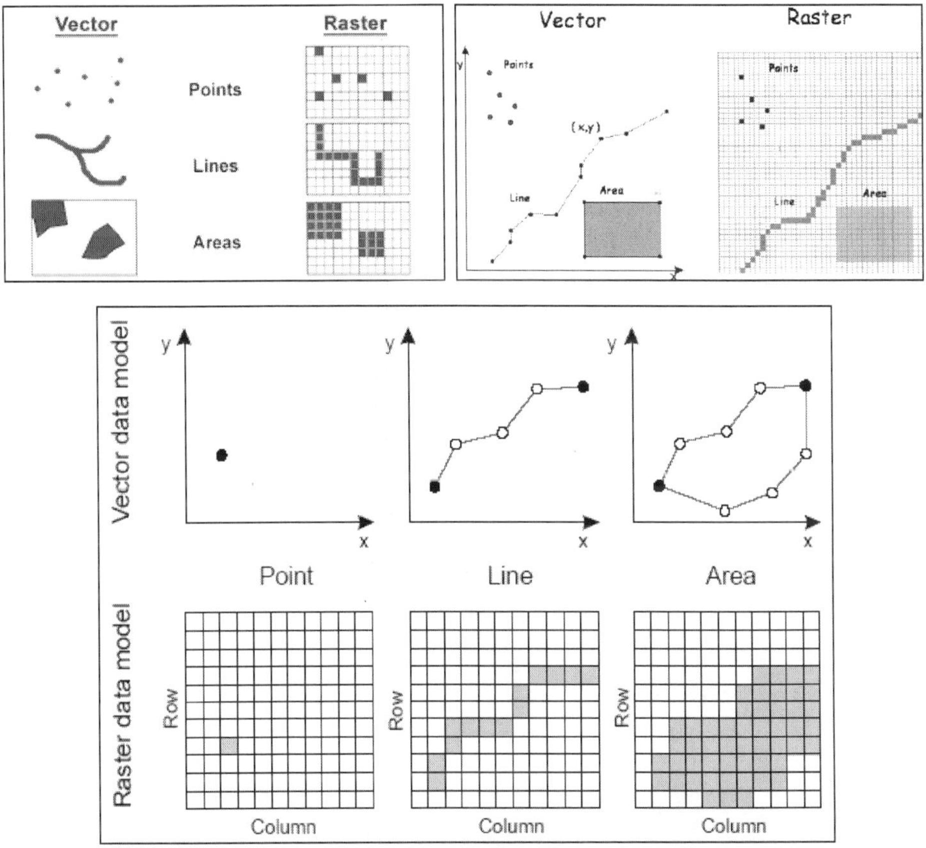

Figure 5.3. Types of Geographic Information System Data.

b) **Vector data**: It is data that has X and Y coordinates assigned to it. Vector files can contain sets of points, lines or polygons that are referenced in a geographic space. Vector data are composed of points, polygon and polylines. e. g. houses are represented by points; roads, rivers, streams are represented by polylines and town are represented by polygons. Some examples of vector file format; Shapefiles, Spatial database engine (ArcSDE), digital line graph (DLG), topologically integrated geographic encoding and referencing (TIGER), vector product format (VPF), national transfer format (NTF), *etc.*

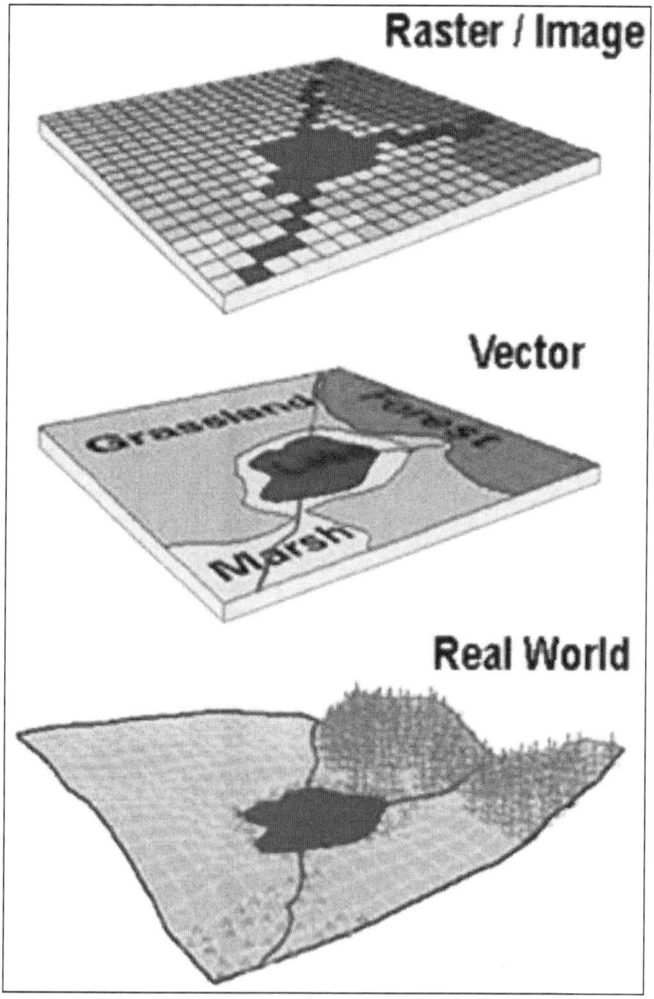

Figure 5.4. Vector and Raster Maps in GIS.

c) **Raster data:** It divides the entire study area into a regular grid of cells. Each cell contains a single value. Raster data is in jpg., tif., gif. or similar format. Items scanned using a flatbed scanner also digital cameras produce same types of files. Some examples of raster file format; Multi-resolution seamless image database (MrSID), network common data form (netCDF), digital raster graphic (DRG), joint photographic experts group (JPEG2000), JPEG file interchange format (JFIF), portable network graphics (PNG), *etc.*

5.6.1 Difference Between Raster Model Data and Vector Model Data

Raster Model Data	*Vector Model Data*
It is simple data structure model	It is more complex data structure
The data structure is less compact	It provides a more compact data structure
Topological relationships are more difficult to represent	Topological relationships are more efficient to represent
High spatial variability is efficiently represented	High spatial variability is inefficiently represented
Raster format is more or less required for efficient manipulation and enhancement of digital images	Manipulation and enhancement of digital images cannot rather than the smooth lines of hand drawn maps.
Overlay operations are easily and efficiently implemented	Overlay operations are more difficult to implemented

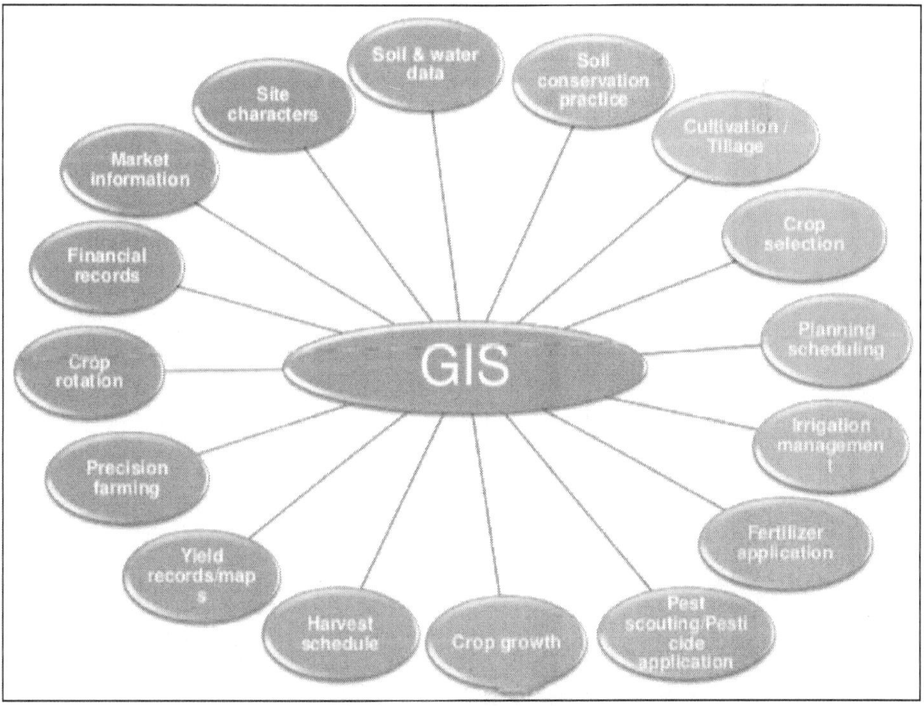

Figure 5.5. Application of GIS in Various Farm Operation.

5.7 Advantages and Disadvantages of GIS

5.7.1 Advantages

a) **GIS helps to improves decision making:** GIS presented detailed and specific information about one or multiple locations which facilitates decision making and makes it easy for the user to make the right choice.

b) **Increases efficiency and reduces cost:** GIS is very cost effective especially given the fact that it is not expensive to maintain in regards to schedules, scheduling time tables or fleet movement and also it is very efficient since it provide users with the specific information about locations.

c) **Makes record keeping easy:** record keeping is very import aspect to geographers it enables them to be able to keep track of change. When using GIS it is easy to record geographical changes this makes it easy for the people responsible of recording to record.

d) **Geographical management:** knowing what will happen in the feature in a geographical space is very important it helps geographer to be able to plan a course of action. GIS helps to manage all these.

e) **Easy to use:** GIS technology is easy to use it can be embedded in to the framework of any organization information system.

f) **Creates employment:** Installing the GIS in to an organizations system means you need some once to run and manage it which means employing people.

5.7.2 Disadvantages

a) Geographic information system (GIS) is very **costly software to acquire**

b) It is very **difficult to create GIS programs** that id user friendly at the same time fast

c) The GIS system typically needs a **complicated command language**

d) For GIS results to be **correct the inputted data must be accurate**, hence the GIS personnel have to be aware of the fact that not all the data are always correct

e) GIS **requires a large amount of data input** and all this data have to be practical for some exercises and hence additional data is inputted

f) For exposed to a large scale, more and **more geographical errors**

g) GIS layers may bring about **big mistakes when it is handled by a non GIS person.**

5.8 Geomatics

Geomatics is defined as the discipline concerned with the collection, distribution, storage, analysis, processing, presentation of geographic data or

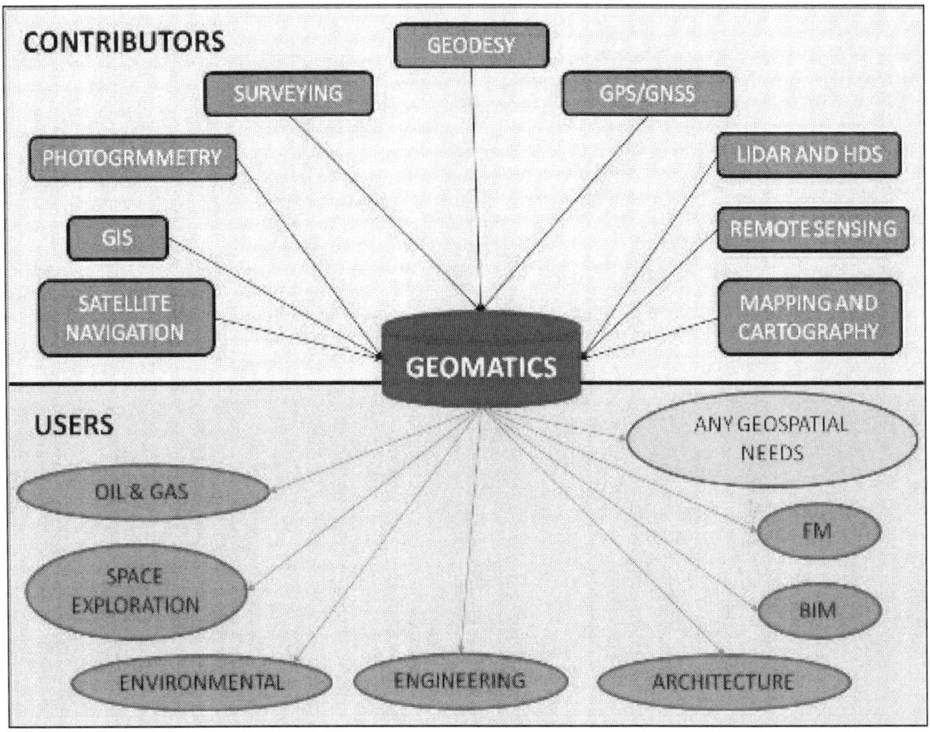

Figure 5.6. Geomatics.

geographic information. Geomatics includes the tools and techniques used in land surveying, remote sensing, cartography, geographic information systems (GIS), global-navigation satellite systems, photogrammetry, geophysics, geography, and related forms of earth mapping.

Chapter 6

Global Positioning System

CHAPTER OUTLINES

6.1 Definition

Global positioning system is a satellite based radio-navigation systems that provides geo-location and time information to a GPS receiver anywhere on or near the earth that helps users to record positional information with an accuracy between 100 to 0.01m.

6.2 Components of GPS

A) Ground control segment/stations

B) Space segment/Satellites

C) User segment/Receivers

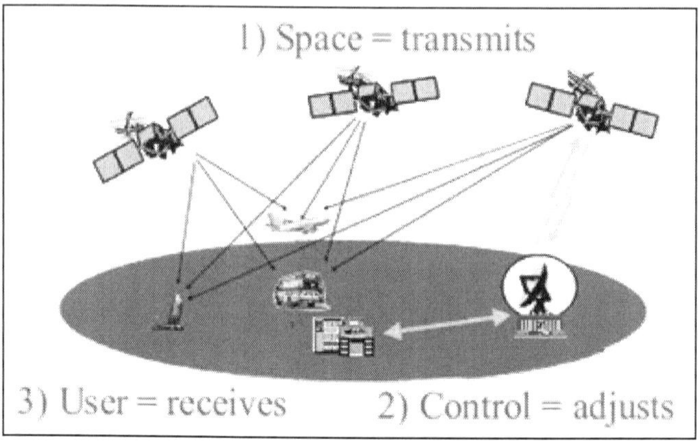

Figure 6.1. Components of GPS.

A) Ground Control Segment

Control Segments formerly consists of 5 tracking stations situated at Hawaii, Ascension Island, Diego Garcia, Kwajalein and the master control facility is located at Schriever air force base (Formerly Falcon AFB) in Colorado springs (Figure 6.2).

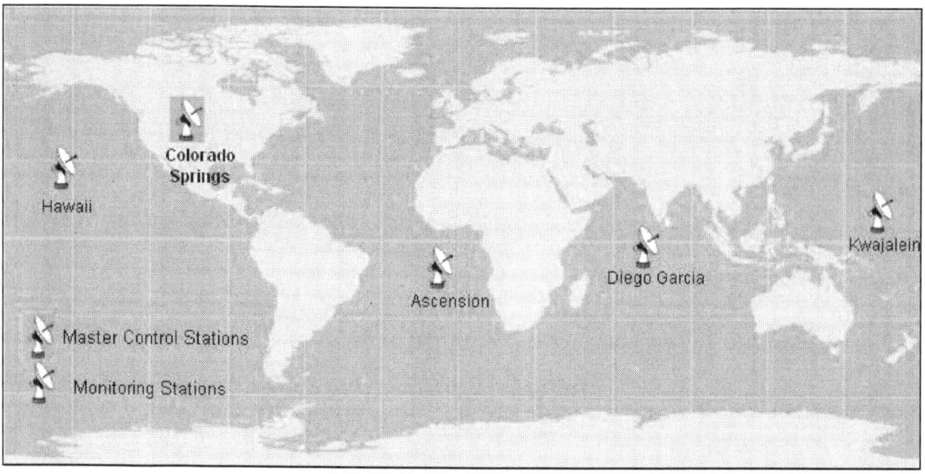

Figure 6.2. Control Segment.

Newly added control stations after 2005 are Washington DC, England, Ecuador, Argentina, Bahrain and Australia (Figure 6.3). These Monitor stations measure signals from the SVs, which are incorporated into orbital models for each satellites. Master stations collect the data about the satellites of this system continuously from the other tracking stations. MCS process the tracking data for computation of satellite ephemerides components and satellite clock parameters. The Master control station uploads ephemeris and clock data to the SVs. The SVs then send subsets of the orbital ephemeris data to GPS receivers over radio signals. The MCS also monitor the position of satellites at any instant of time, the functional capacity of the satellites and variation of the navigation data.

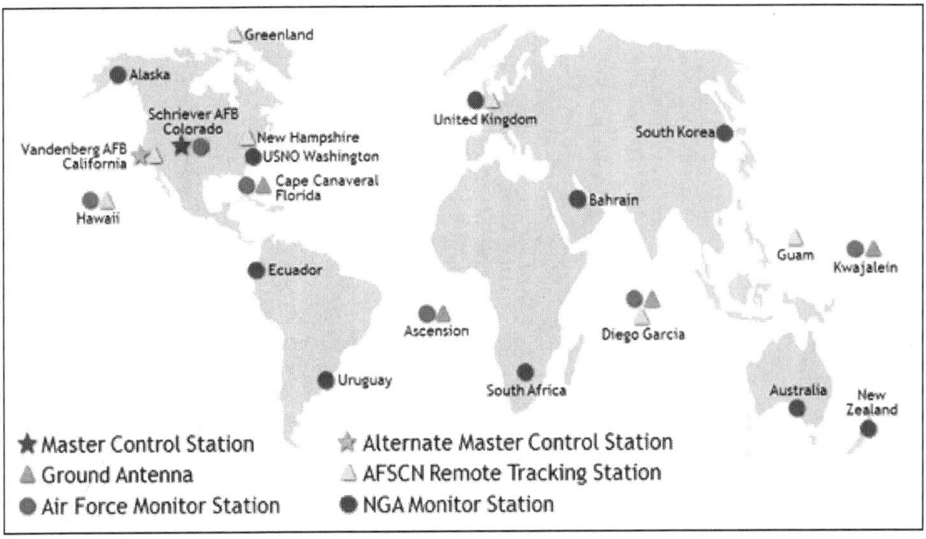

Figure 6.3. GPS Control Segment.

B) Space Segments/Satellites

It Includes satellites and data rockets GPS satellite orbits is circular orbits at 17440 km altitude above earth. Orbital period of 12 (11 hrs 55 min) hrs. The orbital are inclined 55° with respect to equator. The satellites are powered by solar cells to continually orientate themselves to point the solar panels towards the sun and the antennas towards the earth. Each satellite contain four atomic clock, 24 satellite vehicles, 5-8 satellites visible from any points on earth. **Atomic clocks:** That measure time to a high degree of accuracy, time information is placed in the codes broadcast by the satellite so that receiver can continuously determines the time the signal was broadcast.

C) GPS Receiver

The typical receiver is composed of an antenna and pre-amplifier, radio signal microprocessor, control and display device, data recording unit, and power supply. The ground stations send control signals to the GPS satellites. The GPS satellites transmit radio signals and the GPS receivers, receive these signals and use it to calculate its position. when receiving i) One signal - we can only calculate how far away from the satellite we are; ii) two signal - determine two positions where we are; iii) Three signal- determine our exact position on the earth's surface (2D/2) dimensional positioning and iv) when more than three satellites are visible to the GPS receiver it will also calculate the altitude of the receiver (3D/3 dimensional positioning). The GPS satellite constellation is designed to guarantee that at least four satellites are visible from any place on earth at any moment of time. A minimum of four satellites are required to compute the four dimensions of X, Y, Z (position) and Time.

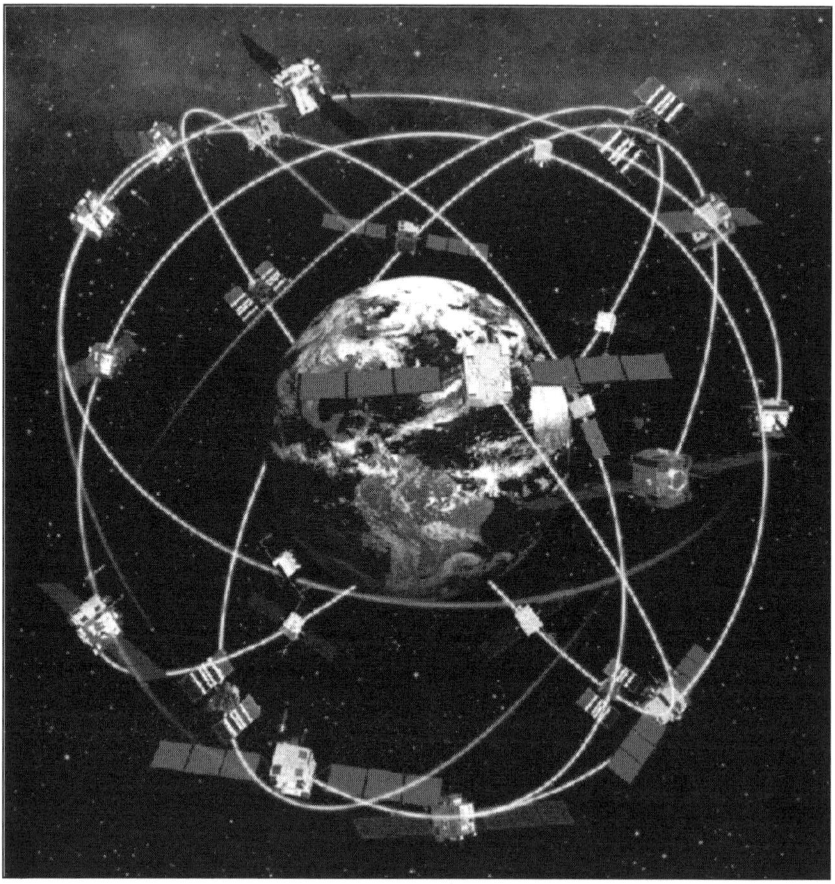

Figure 6.4. GPS Satellite Network (NOAA).

6.3 GPS Working Mechanism

Each of the 24 satellites emits signals to receivers that determine their location or range by computing the difference between the time that a signal is sent and the time it is received. The signal contains data that a receiver uses to compute the locations of the satellites and to make other adjustments needed for accurate positioning. The receiver must account for propagation delays, or decreases in the signal's speed caused by the ionosphere and the troposphere. GPS satellites carry atomic clocks that provide extremely accurate time. The time information is placed in the codes broadcast by the satellite so that a receiver can continuously determine the time the signal was broadcast. With information about the ranges to three satellites and the location of the satellite when the signal was sent, the receiver can compute its own three dimensional position. An atomic clock synchronized to GPS is required in order to compute ranges from these three signals. However, by taking a measurement from a fourth satellite, the receiver avoids the need for an atomic clock. Thus, the receiver uses four satellites to compute latitude, longitude, altitude and time.

Measurement of Distance: Satellites broadcast radio signals (EM radiation)

$$d = v * t$$

where,

d = Distance

v = Velocity (velocity of light 186,000 miles per seconds)

t = Time (difference between send and receive)

6.4 Applications of GPS

a) **Location:** determining a basic position. *e.g.* soil sampling, machinary location

b) **Navigation:** Navigation of civilian ships and planes

c) **Tracking:** monitoring the movement of people, animals and goods

d) **Mapping:** creating maps of the world

e) **Guidance:** Point guidance and swath guidance

f) **Control:** Variable rate application, Variable rate technology

g) **Mapping:** Soil properties, chemical application, chemical prescription, tillage map, topographic, contour mapping, river mapping, planting map, yield and pest mapping

h) **Mining:** track the minerals in various layer of the surface of the earth

i) **Surveying:** Agricultural land survey

Chapter 7

Remote Sensing

CHAPTER OUTLINES

7.1 Definitions

Remote sensing is the process of acquiring information about the earth's surface by measuring its reflected and emitted radiation without coming into direct contact with the object. **or Remote sensing** is the art and science of gathering information about the objects or area of the real world at a distance without coming into direct physical contact with the object under study. **or Remote sensing** is the art and science of obtaining information from a distance, *i.e.* obtaining information about objects or phenomena without being in physical contact with them.

Pisharoth Rama Pisharoty is considered to be the father of remote sensing in India. Pisharoty and his colleagues at Space Application Centre, Ahmedabad introduced remote sensing in India in 1970 with an early detection of coconut wilt disease using colour infrared aerial photographs. **Shunji Murai** is considered to be the father of remote sensing in Asia. The term "remote sensing" first used in the

United States in the 1950s by **Ms. Evelyn Pruitt** of the U.S. Office of Naval Research. The word remote sensing was coined by **Fisher** in 1960.

7.2 History of Remote Sensing

7.2.1 Milestones in History of Remote Sensing

1800 Discovery of infrared by **Sir W. Herchel**

1839 Beginning of practice of photography

1847 Infrared spectrum shown by **J. B. L. Foucault**

1858 Photography from balloons by **Gaspard Tournachon**

1873 Theory of electromagnetic spectrum by **Sir James Cleck Maxwell**

1909 Photography from airplanes by **Wilbur Wright**

1916 World War I: aerial reconnaissance

1935 Development of **radar** in Germany

1940 World War II: Application of non-visible parts of electromagnetic radiation

1959 First space photograph of the earth **(xplorer-6)**

1960 First meteorological satellite launched **(TIROS-I)**

1970 Skylab remote sensing observations from space

1972 Launch LANDSAT-I and rapid advancement in digital image processing

1973 SKYLAB - the American space station was launched

1982 Launch LANDSAT-4 with new generation of sensors (TM)

1986 French commercial earth observational satellite SPOT

1986 Development of hyper-spectral sensors

1990 Developing high resolution space borne systems and first commercial developments in remote sensing

1999 Launch EOS: NASA earth observing mission

1999 Launch of IKONOS, very high spatial resolution sensor systems

7.2.2 Development of Remote Sensing in India

1966 Indian Institute of Remote sensing was established at **Dehradun, Uttarakhand**

1967 First Indian rocket RH-75 launched from Thumba Equatorial Rocket Launching Station near Thruvanathapuram. Also launch of Bhaskar-I and Bhaskar-II

1975 National Remote Sensing Agency (NRSA) established at Hyderabad

1988 March Launch of IRS- 1A (Indian Remote Sensing Satellites) by ISRO

1991 August IRS-IB

1994 October IRS-P2

1994 Developed PSLV (Polar Satellite Launch Vehicle)

1995 December IRS-IC

1996 March IRS-P3

1997 Launched remote sensing satellite from Shriharikota, Hyderabad

7.3 Components and Stages of Remote Sensing

a) **Energy source or Illumination (A)**: It is an energy source for remote sensing which illuminates or provides electromagnetic energy to the target of interest.

b) **Radiation and the Atmosphere (B)**: The energy travels from its source to the target, it will come in contact with and interact with the atmosphere it passes through. This interaction may take place a second time as the energy travels from the target to the sensor.

c) **Interaction with the target (C):** The energy makes its way to the target through the atmosphere, it interacts with the target depending on the properties of both the target and the radiation. The process involved in the interaction of electromagnetic radiation with earth surface are reflection, scattering and transmission.

d) **Recording of energy by the sensor (D):** The energy has been scattered or emitted from the target, we require a sensor (remote-not in contact with the target) to collect and record the electromagnetic radiation.

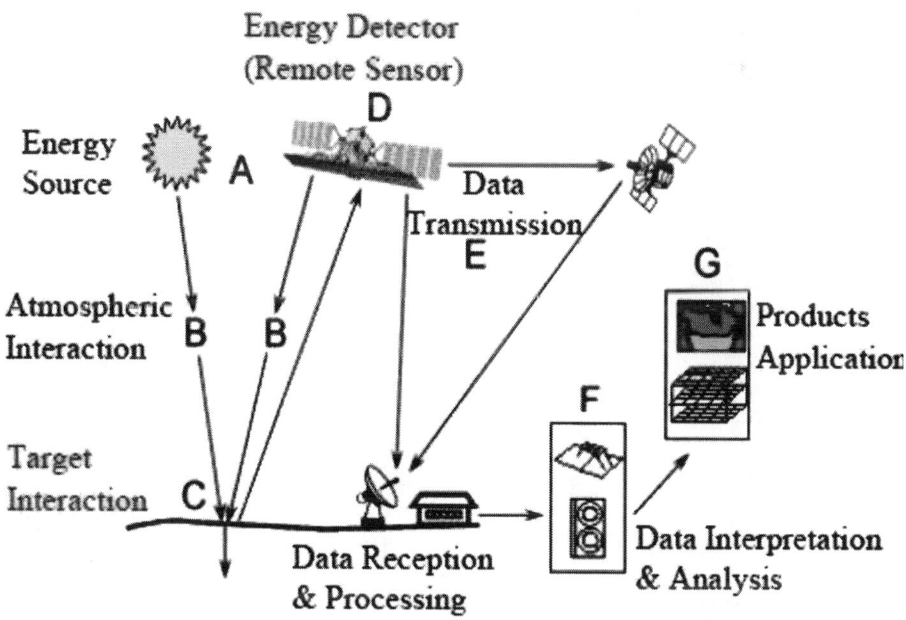

Figure 7.1. Different Components of Remote Sensing.

e) **Transmission, Reception and Processing (E):** The energy recorded by the sensor has to be transmitted, often in electronic form, to a receiving and processing station where the data are processed into an image (hardcopy or digital).

f) **Interpretation and Analysis (F):** The processed image is interpreted, visually or digitally or electronically, to extract information about the target which was illuminated.

g) **Application (G):** The final element of the remote sensing process is achieved when we apply the information we have been able to extract from the imagery about the target in order to better understand it, reveal some new information or assist in solving a particular problem.

7.4 Principle of Remote Sensing

Different objects return different amount and kind of energy in different band of the spectrum incident upon it. This property of the objects depends on the structural, physical, chemical properties, surface roughness, intensity and wavelength of radiant energy reflect or emit different amount of energy in different wavelength ranges of electro-magnetic spectrum. The sensors measures the amount of energy reflected from that object and represents it through an image.

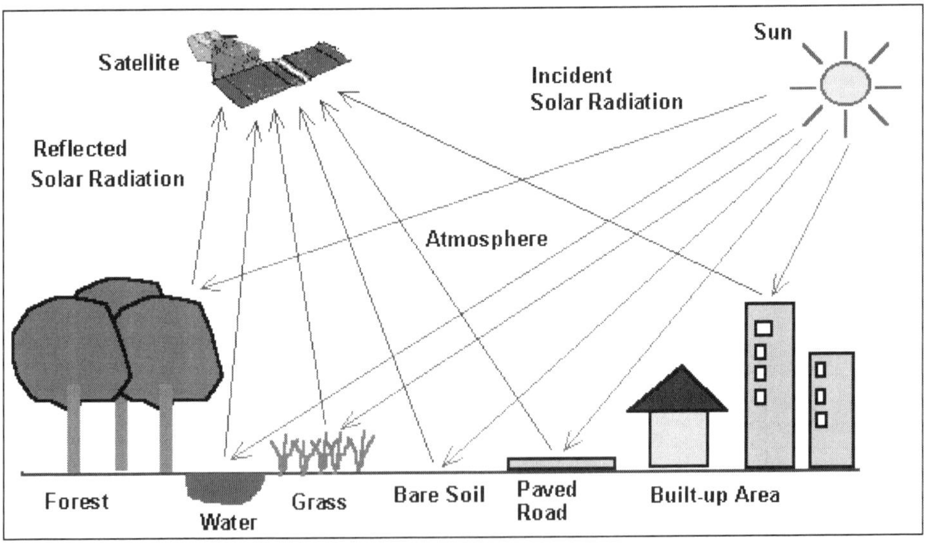

Figure 7.2. Principles of Remote Sensing.

7.5 Types of Remote Sensing

7.5.1 Active Remote Sensing

When remote sensing work is carried out with a man made source of radiations which is used to illuminate a body and to detect the signal reflected form. *e.g.* Radar and Lidar remote sensing.

7.5.2 Passive Remote Sensing

When remote sensing work is carried out with the help of electromagnetic radiations (signals) reflected by a natural body (sun and earth). *e.g.* visible, NIR and Microwave remote sensing.

7.6 Applications of Remote Sensing in Agriculture

a) **Crop production forecasting:** Remote sensing is used to forecast the expected crop production and yield over a given area and determine how much of the crop will be harvested under specific conditions. Researchers can be able to predict the quantity of crop in a given farmland over a given period.

b) **Water resources mapping:** Remote sensing is instrumental in the mapping of water resources that can be used for agriculture over a given farmland. Through remote sensing, farmers can tell where water resources are available for use over a given land and whether the resources are adequate.

c) **Soil mapping:** Through soil mapping, farmers are able to tell which soils are ideal for which crops and which soil require irrigation and which ones do not. This information helps in precision agriculture. Soil maps are another type of maps developed using remote sensing data. These maps can be compiled on the basis of airborne or satellite images acquired when the degree of soil coverage by plants is less than 30-50 per cent. Soil maps present homogeneous soil zones with similar properties and conditions for plant growth. These maps are useful in determining soil sampling locations for detailed studies of soil, soil moisture sensors location or developing irrigation plans.

d) **Monitoring of droughts:** Remote sensing technology is used to monitor the weather pattern of a given area. The technology also monitors drought patterns of the area too. The information can be used to predict the rainfall patterns of an area and also tell the time difference between the current rainfall and the next rainfall which helps to keep track of the drought.

e) **Soil moisture estimation:** Remote sensing gives the soil moisture data and helps in determining the quantity of moisture in the soil and hence the type of crop that can be grown in the soil.

f) **Crop identification:** Remote sensing has played an important role in crop identification especially in cases where the crop under observation shows some mysterious characteristics. To identify a particular crop, we know about its growth cycle (germination, growth, pollination, senescence) and how the crops reflect the near-infrared at each of their various growth stages. Using the different near infrared reflectance we have to discriminate between two crops. Having the knowledge of when each crop is planted and harvested, we can estimate the percentage of vegetation cover through the growth period, assuming no external factors (stress, disease *etc.*) affect its growth. By using multi-date data from one growing period, it is possible to identify the different crop types, because the vegetation cover of each crop changes at different rates. By combining this information with remote sensing data, we can discriminate between different crops and also identify them. This information serves to predict grain crop yield, collecting crop production statistics, facilitating crop rotation records, mapping soil productivity, identification of factors influencing crop stress, assessment of crop damage and monitoring farming activity.

g) **Crop acreage estimation:** Remote sensing has also played a very important role in the estimation of the farmland on which a crop has been planted. This is usually a cumbersome procedure if it is carried out manually because of the vast sizes of the lands being estimated.

h) **Crop yield modeling and estimation:** Remote sensing also allows farmers and experts to predict the expected crop yield from a given farmland by estimating the quality of the crop and the extent of the farmland. This is then used to determine the overall expected yield of the crop. The particular manner vegetation reflects the electromagnetic radiations hence, we can assess the crop status by using remote sensing data. By combining these data with additional data (climatic conditions) in complex models, it is possible to estimate the final yield of a crop field at a very early stage. Remote sensing data and the derived vegetation indices can be used to make estimates of crop yield, provided that a relationship between the index or indices and crop yield for the particular crop in the specific location has been established in the past.

i) **Yield maps:** Spatial distribution of plant biomass and bio-physical parameters within fields is not even, but very patchy, defined by soil conditions, therefore it is possible to compile yield maps. Such maps created on the basis of satellite images acquired in many seasons represent the spatial variability in crops yield regardless of plant species. Yield maps are applied to determine patterns of fertilization and irrigation and can be used in planning programs for the eradication of weeds, pests and plant diseases. In such programs, fertilizer or pesticides doses are adopted to the productivity of a specific spot on the field.

j) **Identification of pests and disease infestation:** Remote sensing technology a plays a significant role in identification of pests in farmland and gives data on the right pests control mechanism to get rid of the pests and diseases on the farm.

k) **Assessment of crop damage and crop progress:** In the event of crop damage or crop progress, remote sensing technology can be used to penetrate the farmland and determine exactly how much of a given crop has been damaged and the progress of the remaining crop in the farm.

l) **Identifying stressed plants:** Chlorophyll is an essential part of the process of photosynthesis. It absorbs solar energy to provide power for the process of photosynthesis. Because it absorbs energy, it has an important effect on the amount of energy that is reflected. With remote sensing we can directly estimate how much chlorophyll there is in a plant. By combining more than one bands of the recorded remote sensing data, we can create vegetation indices and use them to estimate crop status. Depending on the visible and near-infrared reflectance, the produced vegetation indices give us an indication on the amount of chlorophyll present in the plants. With this information we can estimate how much stress the plants are under.

m) **Detection, diagnosis and control of plant diseases:** By combining agricultural knowledge with remotely sensed data, it is possible to have early warning and prevent a pest or a disease from affecting the crops, by taking appropriate action at an early stage. **Hyper-spectral remote sensing** helps to detect crops diseases at early stages and facilitates the targeted delivery of the necessary treatments. It is possible to assess the extent of the damage caused by pests and diseases. The symptoms of such attacks usually cause the break-down of chlorophyll and we can identify the reduction of chlorophyll concentration in the plants through remote sensing. In addition to loss of chlorophyll, pest and diseases can cause the destruction of whole leaves. This leads to a reduction in the total leaf area and as a result, the reduction of the plant's capacity for photosynthesis. As we are able to estimate the Leaf Area Index (LAI) of a group of plants, it is possible to identify an insect attack at an early stage and warn the farmers to take the appropriate measures.

n) **Plant breeding research:** Plant breeders and other researchers employ many different scientific methods in their traditional or molecular-based programs, ranging from direct measurement of grain or seed production to more complex marker-assisted techniques. Near Infrared (NIR) spectroscopy provides the agricultural researcher with a non-destructive tool for plant, seed and other agricultural product analysis. Once calibration models have been developed a single sample can be measured in less than one minute.

o) **Soil analysis:** For decades, scientists have used high-resolution reflectance spectra of minerals and soils to determine soil mineralogy and to assess soil physical properties. A major breakthrough in these studies has been the use of visible-near infrared spectroscopy to develop quantitative calibrations for rapid characterization of soil nutrients and various physical properties of soils. The coupling of this technology with remote sensing data, geo-referenced ground surveys and new spatial statistical methods has resulted in the improved capability for large area soil assessments.

p) **Land cover mapping:** Land cover corresponds to the physical condition of the ground surface (e. g. forest, grassland, concrete pavement *etc.*), while land use reflects human activities such as the use of the land (e. g. industrial zones, residential zones, agricultural fields *etc.*). Information on land cover and changing land cover patterns is directly useful for determining and implementing environment policy and can be used with other data to make complex assessments (*e.g.* mapping erosion risks). Land cover change detection is necessary for updating land cover maps and the management of natural resources.

q) **Identification of planting and harvesting dates:** Remote sensing is used to observe a variety of factors including the weather patterns and the soil types to predict the planting and harvesting seasons of each crop.

r) **Irrigation monitoring and management:** Remote sensing gives information on the moisture quantity of soils. This information is used to determine whether a particular soil is moisture deficient or not and helps in planning the irrigation needs of the soil.

s) **Identification of problematic soils:** Remote sensing has also played a very important role in the identification of problematic soils that have a problem in sustaining optimum crop yield throughout a planting season.

t) **Determination of water content of field crops:** Remote sensing also plays an important role in the estimation of the water content in the field crops.

u) **Flood mapping and monitoring:** Using remote sensing technology, farmers and agricultural experts can be able to map out the areas that are likely to be hit by floods and the areas that lack proper drainage. This data can then be used to avert any flood disaster in future.

v) **Collection of past and current weather data:** Remote sensing technology is ideal for collection and storing of past and current weather data which can be used for future decision making and prediction.

w) **Crop intensification:** Remote sensing can be used for crop intensification that includes collection of important crop data such as the cropping pattern, crop rotation needs and crop diversity over a given soil.

x) **Precision farming:** Remote sensing has played a very vital role in precision agriculture. Precision agriculture has resulted in the cultivation of healthy crops that guarantees farmers optimum harvests over a given period of time.

y) **Climate change monitoring:** Remote sensing technology is important in monitoring of climate change and keeping track of the climatic conditions which play an important role in the determination of what crops can be grown where.

z) **Soil management practices:** Remote sensing technology is important in the determination of soil management practices based on the data collected from the farms.

aa) **Land mapping:** Remote sensing helps in mapping land for use for various purposes such as crop growing and landscaping. The mapping technology used helps in precision agriculture where specific land soils are used for specific purposes.

Chapter 8

Crop Simulation Model and their Uses

CHAPTER OUTLINES

8.1 Definitions

Simulation is the process of building models and analyzing the system. **or Simulation** is the process in which a artificial mathematical substitute system is developed to represent and function as a real system.

Model is an equation or set of equations which represent the behavior of a system. **or Model** is a simplified representation of a system or a process.

Modeling is the use of equations or sets of equations to represent the behavior of a system.

Crop model is the simple representation of a crop. Crop model is tools of systems research which helps in solving problems related to crop production.

8.2 Types of Models

Depending upon the purpose for which it is designed the models are classified into different groups:

a) **Statistical model:** These models express the relationship between yield or yield components and weather parameters. In these models relationships are measured in a system using stastical techniques. The model does not explain the mechanism, approach and aspect of weather parameters by which they influence yield. The model do not take soil variability, genetic potential and management into account.

b) **Mechanistic model:** These models explain not only the relationship between weather parameters and yield, but also explain the mechanism of influencing dependant variables. These models are based on physical selection.

c) **Deterministic model:** These models estimate the exact value of the yield or dependent variables. These models also have defined coefficients.

d) **Stochastic model:** These models use the value of weather parameters at some or other probability level. Hence, output components are estimated within a range depending upon probability level of dependent variables. These models define or state of dependant variable at a given rate.

e) **Dynamic model:** Time is included as a variable. Both dependent and independent variable are having value which remain constant over a given period of time.

f) **Static model:** Time is not included as a variable. Both dependent and independent variable are having value which remain constant over a given period of time.

g) **Descriptive model:** These models is helpful to determine quickly the weight of the crop where no observation was made.

h) **Explanatory model:** These models consists of quantitative description of the mechanisms and process that cause the behavior of the system. It contains description of distinct processes such as leaf area expansion and tiller production.

i) **Simulation model:** These models use one or more sets of different equation and calculate both rate and state variables over time. It used to estimate agricultural production as a function of weather, soil condition and crop management.

8.3 Crop Simulation Model

Crop Simulation Model (CSM) is a simulation model that describes processes of crop growth and development as a function of weather conditions, soil conditions and crop management.

8.3.1 Crop Simulation Model

Input Data Required	*Model*	*Output*
Weather Data Crop Data Soil Data Crop Management Data Pest Data	⇒ Crop Simulation model ⇒	Daily biomass accumulation (Growth, Yield)

A) Input Data Requirement

a) **Weather Data:** Maximum temperature, minimum temperature, rainfall, relative humidity, solar radiation, wind speed. It required at daily time steps to assess daily crop growth process.

b) **Crop Data:** Crop name, crop variety, crop phonology, leaf area index, grain yield and test weight of grains.

c) **Soil Data:** Thickness of soil, pH, electric conductivity, nitrogen, phosphors, potash, organic matter, organic carbon, texture, moisture, Field capacity, PWP and Bulk density.

d) **Crop Management Data:** Sowing date, seed rate, depth, irrigation, fertilizers, manures, crop residue, date of application, time of application and C: N ratio.

e) **Pest Data:** Name and type of the pest, mode of attack and pest population.

8.3.2 Commonly Used Crop Simulation Models

a) DSSAT (Decision Support System for Agro-technology Transfer) Model

It is a software application which has crop simulation model for different crop. The input data to this software is daily weather data, soil surface and profile information and detailed crop management with output as experimental data so that farmers can compare the predicted and actual output. With the help of DSSAT farmers can avail the information on crops and new farm practices to achieve the high crop yield.

To meet the requirements of resource poor farmers in the tropics and sub tropics IBSNAT (International Benchmark Sites Network for Agro-technology Transfer) began in 1982. This was under a contract from the U.S. Agency for International Development to the University of Hawaii at Manoa, USA. The major product of IBSNAT was a Decision Support System for Agro Technology Transfer (DSSAT). The network members lead by J.W. Jones, Gainesville, USA developed this. The DSSAT is being used as a research and teaching tool. As a research tool its role to derive recommendations concerning crop management and to investigate environmental and sustainability issues is unparalleled. The DSSAT products enable users to match the biological requirements of crops to the physical characteristics of land to provide them with management options for improved land use planning. The DSSAT is being used as a business tool to enhance profitability and to improve input marketing.

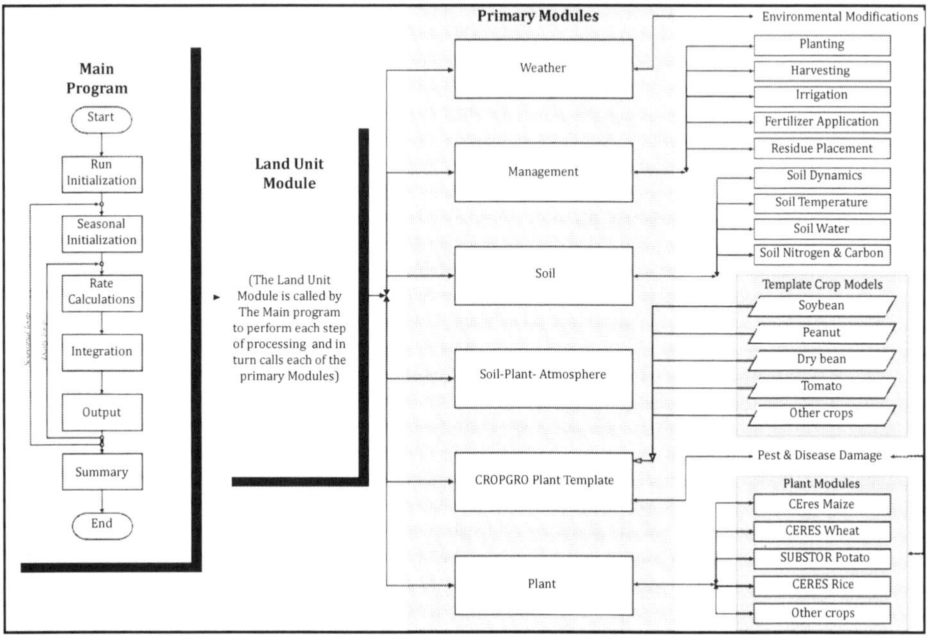

Figure 8.1. DSSAT Model.

b) APSIM (Agricultural Production Systems Simulator) Model

It was created by CSIRO, Canberra, Australia. Required input data of soil properties, daily climate data, cultivar characteristics, and agronomic management and Outputs are changes in crop and pasture yields, yield components, soil erosion losses, for different climate change scenarios. The APSIM is an effective tool for analyzing whole farm systems, including crop and pasture sequences and rotations, and for considering strategic and tactical planning. APSIM allows users to improve understanding of the impact of climate, soil types, and management on crop and pasture production. It is a powerful tool for exploring agronomic adaptations such as changes in planting dates, cultivar types, fertilizer/irrigation management, *etc.*

c) InfoCrop Model

It is the crop model developed by IARI, New Delhi. It is dynamic simulation model for the assessment of crop yields, losses due to pests, and environmental impact of agro-ecosystems in tropical environments. InfoCrop is a process based dynamic simulation model for simulating growth, development and yield of rice, wheat, maize, sorghum, pearl millet, mustard, soybean, chick pea, pigeon pea,

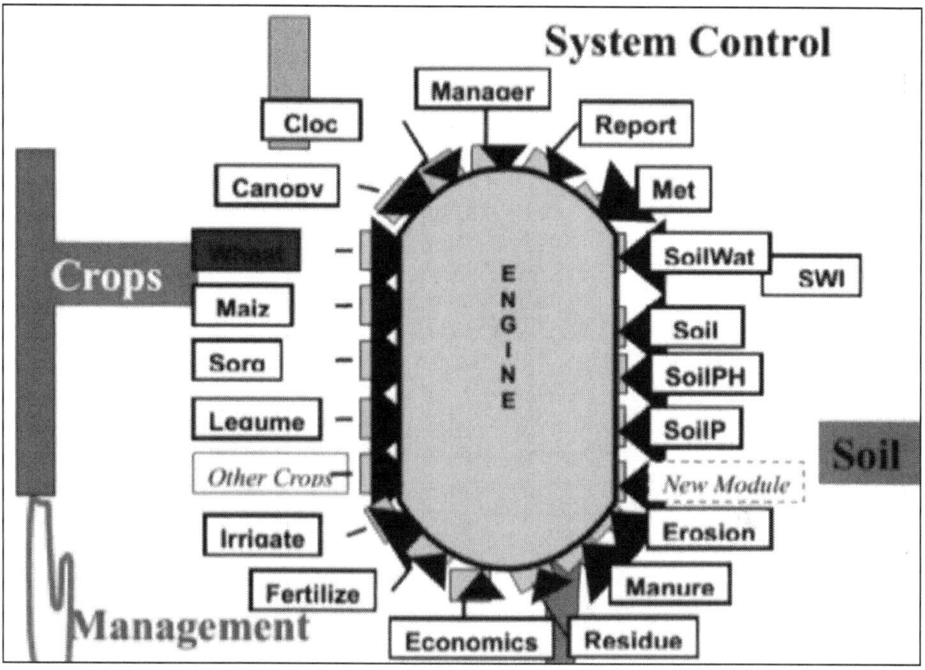

Figure 8.2. APSIM Model.

potato and cotton. It simulates the effects of weather, soil and crop management (sowing, seed rate, organic matter nitrogen and irrigation) and pests. It provides daily and summary outputs on various growth and yield parameters, nitrogen uptake, green house gas emissions, soil water and nitrogen balance. It is used for several applications including yield forecast and climate change studies. It is known to perform better for tropical regions.

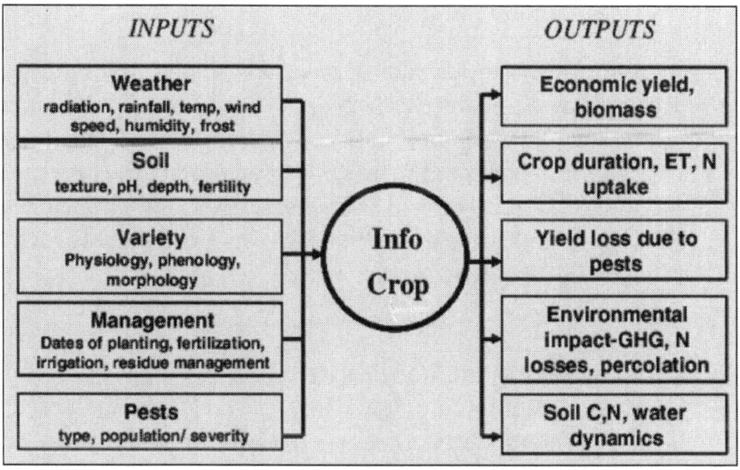

Figure 8.3. InfoCrop Model.

d) AquaCrop Model

It is the crop growth model developed by FAO to address food security and assess the effect of the environment and management on crop production. Aqua Crop simulates the yield response of herbaceous crops to water and is particularly well suited to conditions in which water is a key limiting factor in crop production.

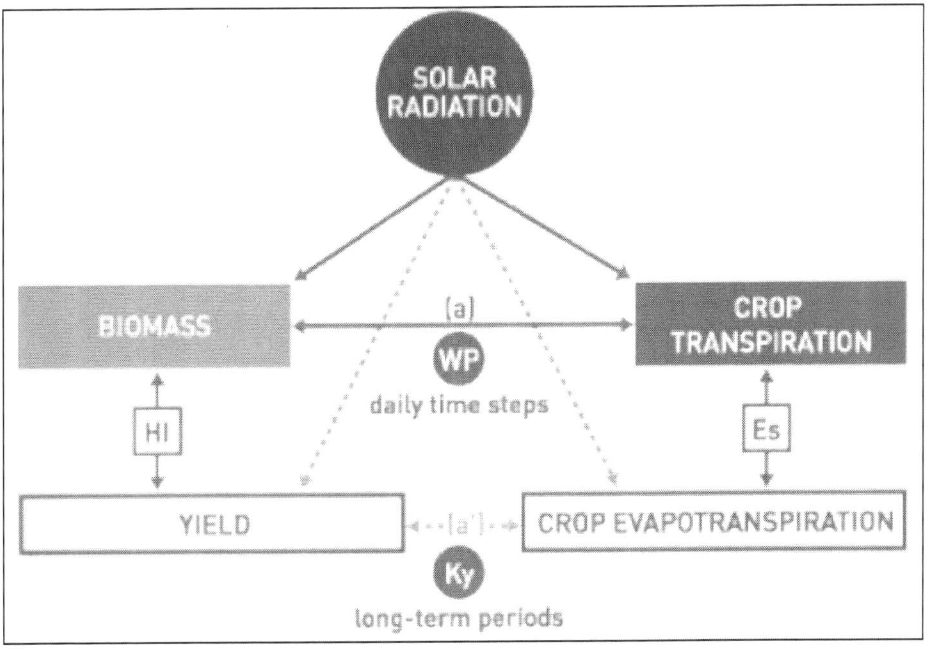

Figure 8.4. Components of AquaCrop Model.

e) CERES (Crop Estimation through Resource and Environment Synthesis) Model

This model require the input data of weather *i.e.* minimum and maximum air temperature, longitude and latitude *etc.* data of soil *i.e.* soil water content soil texture *etc.* and data of management practices *i.e.* sowing dates, irrigation date *etc.* the model takes the input data mentioned above and predict the different options of management practices for crop yield estimation. It also forecast the crop status during growing stages. There are different CERES model available for crops like CERES-wheat model, CERES –Rice model.

f) CropSyst Model

It is a multi-year multi-crop daily time-step crop simulation model being developed by a team at Washington State University's Department of Biological Systems Engineering. The model has been developed to serve as an analytic tool to study the effect of cropping systems management on productivity and the environment. The model simulates the soil water budget, soil-plant nitrogen budget, crop canopy and root growth, dry matter production, yield, residue production and decomposition, and erosion. Management options include: cultivar selection, crop rotation, irrigation, nitrogen fertilization, tillage operations and residue management.

g) EPIC (Environmental Policy Integrated Climate) Model

The model was designed for estimating the productivity of soil caused by soil erosion. This model forecast the impact of different management strategies on soil, water, nutrient and pesticide on soil loss, quality of water and yield for area with same types of soil and management.

h) WOFOST Model

This model is designed by World food studies in Wageningen (Netherlands). This model require the rain, temp, wind speed, air humidity, soil moisture content, wilting point and management practices as a input. The model analyzes the consequences of weather change, different sowing approaches and crop yield variation for different types of soil. It also use for prediction of the crop yield.

8.3.3 Steps Involved in Building a Crop Simulation Model

Defining the purpose of the model

⇩

Define system and its boundaries

⇩

Define key variables in system

⇩

Developing the structure of the model

⇩

Quantify relationship using different mathematical equations and function

⇩

Calibration

⇩

Validation

⇩

Sensitivity analysis

⇩

Simplification

⇩

Use of models in decision support

8.3.4 Limitations of Modeling

a) Required large amount of input data, which may not be available

b) Required skilled manpower, good knowledge of computers and computer language

c) All the existing complexity of biological systems not take into account

d) Models developed for a specific region cannot be used as such in another region

e) Improve efficiency of field experiments, but they do not eliminate the need for continued experimentation

f) Proper parameterization and calibration is needed before using a model

8.4 Uses of Crop Simulation Model

a) Yield forecasting: Prediction of crop yield before harvesting depending upon the environmental condition Yield forecasting for industries over large areas is important to the producer (harvesting and transport), the processing agent (milling period) as well as the marketing agency. The technique uses weather records together with forecast data to estimate yield across the industry.

b) Impact assessment of climatic variability and climatic change: by using these models we can find the impact of climate variation on crop growth.

c) Use in agro-met advisory system (AAS): Crop simulation model can be used as a decision making tool in the Agro-met Advisory System of the country.

d) Analyze the actual crop yield and predicted crop yield by using crop simulation models.

e) Field specific experimentation: Crop simulation model will reduce the field visit.

f) Policy analysis: It will help policy makers to frame better policies s by analyzing the result of these crop models.

g) Crop management optimization: By improving management practices production of crop will improve. Optimizing management likes dates of planting, variety, irrigation and nitrogen fertilizer.

h) Study of responses to planting date: Decide optimum sowing date for maximum yield.

i) Plant breeding: Simulation models use the genetic characters in the foam of rate coefficient or constant in crop growth. Thus, breeding work can be undertake to use only useful characters to produce high yielding varieties, insect and disease resistant cultivars capable to compete with yield.

j) **Study of insect and disease development:** Models can be used to develop relationship between weather condition and insect pest occurrence and disease development. Such relationship can be used to forecast the events of pest occurrence and disease occurrence so as to undertake plant protection programme at suitable time to avoid yield losses and economic losses.

k) **Yield gap analysis:** Model predicted decrease or increase in yield hence, used to quantify the yield gap between actual and potential yield.

l) **Evaluation of potential yield:** Model predicted the yield hence, used to evaluate the potential yield.

m) **Integration of knowledge across disciplines:** Adoption of a modular approach in model coding allows the scientist to pursue his discipline-oriented research in an independent manner and at a later stage to integrate the acquired knowledge into a model. For example, the modular aspect of the APSIM software allows the integration of knowledge across crops as well as across disciplines for a particular crop. Adoption of a modular framework also allows for the integration of basic research that is carried out in different regions, countries and continents. This ensures a reduction of research costs (*e.g.*, through a reduction in duplication of research) as well as the collaboration between researchers at an international level.

n) **Improvement in experiment documentation and data organization:** Simulation model development, testing and application demand the use of a large amount of technical and observational data supplied in given units and in a particular order. Data handling forces the modeler to resort to formal data organization and database systems. The systematic organization of data enhances the efficiency of data manipulation in other research areas (*e.g.*, productivity analysis, change in soil fertility status over time).

o) **Genetic improvement:** As simulation models become more detailed and mechanistic, they can mimic the system more closely. More precise information can be obtained regarding the impact of different genetic traits on economic yields and these can be integrated in genetic improvement programs, *e.g.* the NTKenaf model. Researchers used the modeling approach to design crop ideotypes for specific environments.

p) **Risks assessment and investment support:** Using a combination of simulated yields and gross margins, economic risks and weather-related variability can be assessed. These data can then be used as an investment decision support tool.

q) **Global climate change and crop production:** Increased levels of CO_2 and other greenhouse gases are contributing to global warming with associated changes in rainfall pattern. Assessing the effects of these changes on crop yield is important at the producer as well as at the government level for planning purposes.

r) **Precision Farming:** an interface between PC ARC/INFO GIS and SIMPOTATO simulation model to study potato yield and N leaching distribution for site-specific crop management (precision farming) in a field.

s) **Introduction of a new crop**: Simulation model based on scientific data collected few data sets in the new environment helps in the assessment of temporal variability in yield using long term climatic data. Running the simulations with meteorological data in a balanced network of locations also helps in locating the industry.

t) **Site specific farming:** Profit maximization may be achieved by managing farms as sets of sub-units and providing the required inputs at the optimum level to match variation in soil properties across the farm.

Chapter 9

Soil Test Crop Response

CHAPTER OUTLINES

9.1 Definition

STCR approach is aiming at obtaining a basis for precise quantitative adjustment of fertilizer doses under varying soil test values and response conditions of the farmers and for targeted levels of crop production.

ICAR established the AICRP on STCR in **1967** and the STCR concept was developed by **Ramamoorthy**. It provides the relationship between a soil test value and crop yield.

9.2 Objectives of STCR

a) To develop soil test based fertilizer prescription under integrated plant nutrition system (STCR-IPNS) for various crops and soils

b) To study relationship between soil test values for available N, P, K and yield response to important crops

c) To derive yield targeting equation of important crops for making fertilizer recommendations

d) To evaluate the extent to which fertilizer needs of crops can be reduced in relation to conjunctive use of organic manures

e) To evolve a basis for fertilizer recommendation for a whole cropping sequence based on initial soil test values

f) To evaluate the joint use of chemical fertilizers and organic manures for enhanced nutrient use efficiency

9.3 Soil Analysis and Soil Testing

9.3.1 Soil Analysis

Soil analysis is a set of various chemical processes that determine the amount of available plant nutrients in the soil, but also the chemical, physical and biological soil properties important for plant nutrition or soil health. Soil analysis imply; taking soil samples, laboratory analysis of samples and the interpretation of the results by the issuance of fertilizer recommendation. The aims of soil analysis are;

a) To determine the level of availability of nutrients

b) To predict the increase in yields and profitability of fertilization

c) To provide the basis for calculating the required fertilizing of each crop

d) To evaluate the status of each nutrient element and simultaneously determine the compensation plan (nutrient management).

According to the content of nutrients in the soil obtained by chemical analysis and the requirements of the culture for a certain yield, it is easier to determine the amount of required fertilizer to achieve high and quality yields. High quality soil analysis is basis of planning of fertilizing, and thus the quality of the entire production cycle, which results in a high quality and yield and better farm management. **Soil fertility** is the ability of soil to provide all essential plant nutrient in available form; **soil productivity** is the ability of soil to produce crops and **soil fertility evaluation** is the measurement of nutrient supplying capacity of a soil.

9.3.2 Soil Testing

Soil testing is an important tool in high yield farming. Soil testing is diagnostic tool for determining the nutrient needs of plants and for environmental assessments. The aims of soil testing are:

a) To asses fertility status of soil

b) To predict the amount of fertilizer needed

c) To estimate fertilizer requirement of an area
d) To compare fertility level of two or more areas

Soil testing consisting of four phases; soil sampling, sample analysis, data interpretation and soil management recommendations. Soil tests can help to determine soil fertility levels and identify nutrient deficiencies, potential toxicities and trace minerals. They are also important for monitoring the stages of land degradation and acting as a first step in defence by regularly gathering information on which to base land management and fertiliser decisions over time.

9.4 Approaches for Soil Test Based Fertilizer Recommendation to Crops

a) General recommendations approache
b) Soil fertility rating approache
c) Critical limit approache
d) Fertility gradient approach
e) Recommendations for cropping sequence
f) Integrated soil testing approach
g) Targeted yield approach: It is developed by Ramamoorthy. It combine of; a) Calculate the amount of nutrients removed by a crop for certain yield level, b) Calculate contribution made by the available nutrients from soil and fertilizer and c) Calculate the amount of fertilizer to be added to meet the requirement of crop for desirable yield.
h) $\text{Nutrient index} = \frac{(NL \times 1) + (NM \times 2) + (NH \times 3)}{NT}$

where,

NL, NM, NH = No. of soil sample falling in low, medium and high categories

NT = Total number of soil sample analyzed

According to Parker, 1951; NR < 1.5 = fertility status is low in that area, NR between 1.5 to 2.5 then fertility status is medium in that area and NR > 2.5 = fertility status is high in that area.

According to Rammoorty and Bajaj, 1969; NR < 1.67 = fertility status is low in that area, NR between 1.67 to 2.33 = fertility status is medium in that area and NR > 2.33 = fertility status is high in that area

9.5 Parameters of STCR Field Experiments

Targeted yield approach followed in soil test crop response field experiments

a. Nutrient requirement in kg for producing one quintal of grain
= Total uptake of nutrient Grain+Straw in (kg/ha)/Yield of grain (q/ha0

b. Contribution from soil (CS per cent) = Total uptake of nutrient in control plot (kg/ha)/Available soil test values of control plots (kg/ha) x 100
c. Contribution from fertilizer (CF per cent) = Total uptake of nutrient in treated plots/Available soil test values of nutrient treated plots x cs
d. Contribution per cent from fertilizer = C.F./Fertilizer dose applied in (kg/ha) x 100

9.6 Yield Targeting Equation

$$F\,nutrient = \left(\frac{NR}{CF/100}\right) \times T - \left(\frac{CS}{CF}\right) \times S\,nutrient$$

where,

F nutrient = Fertilizer nutrient to be applied (kg/ha)

T = Targeted yield (q/ha)

S nutrient = Soil available nutrient (kg/ha)

NR = Nutrient requirement kg/ha of grain production

CF = Contribution from fertilizer

CS = Contribution from soil

9.6.1 Yield Targeting Equation for different Crops by MPKV, Rahuri (Maharashtra)

Since 1968, soil test response correlation project, at MPKV, Rahuri has been working on targeted yield approach. Now, it is possible to make targeted yield fertilizer recommendation. These recommendation indicates the amount of nutrients from soil and fertilizer required for specified yield.

a) **Wheat** (HD-2189) FN = 7.54 T - 0.74 SN
FP_2O_5 = 1.90 T - 2.88 SP
FK_2O = 2.49 T - 0.22 SK

b) **Rice** Transplanted (Indrayani) FN = 5.20 T - 0.34 SN
FP_2O_5 = 9.40 T - 13.66 SP
FK_2O = 2.73 T - 0.16 SK

c) **Rice** Upland (R-24) FN = 5.52 T - 0.54 SN
FP_2O_5 = 2.19 T - 0.83 SP
FK_2O = 2.37 T - 0.05 SK

d) **Sugarcane** (Seasonal- Co 86032) FN = 4.76 T - 1.34 SN
FP_2O_5 = 1.24 T - 1.55 SP
FK_2O = 2.73 T - 0.21 SK

e) **Sugarcane** (Adsali Co-7219)
$FN = 4.39\ T - 1.56\ SN$
$FP_2O_5 = 1.62\ T - 4.56\ SP$
$FK_2O = 1.86\ T - 0.37\ SK$

f) **Cotton** (RHR-253)
$FN = 13.1\ T - 0.75\ SN$
$FP_2O_5 = 6.83\ T - 2.84\ SP$
$FK_2O = 8.57\ T - 0.18\ SK$

g) **Groundnut** (Kharif-Phule Pragati-JL-24)
$FN = 4.16\ T - 0.37\ SN$
$FP_2O_5 = 4.96\ T - 4.36\ SP$
$FK_2O = 3.14\ T - 0.16\ SK$

h) **Groundnut** (Summer- SB-IX)
$FN = 4.18\ T - 0.40\ SN$
$FP_2O_5 = 8.23\ T - 6.15\ SP$
$FK_2O = 3.22\ T - 0.10\ SK$

i) **Sorghum** (Kharif- CSH-9)
$FN = 4.58\ T - 0.96\ SN$
$FP_2O_5 = 2.21\ T - 6.94\ SP$
$FK_2O = 3.34\ T - 0.22\ SK$

j) **Sorghum** (Rabi- CSH-8 R)
$FN = 4.7\ T - 0.77\ SN$
$FP_2O_5 = 2.0\ T - 4.29\ SP$
$FK_2O = 3.35\ T - 0.33\ SK$

k) **Pearl millet** (Saburi-RRBH-8609)
$FN = 3.31\ T - 0.38\ SN$
$FP_2O_5 = 3.38\ T - 4.11\ SP$
$FK_2O = 1.65\ T - 0.06\ SK$

l) **Finger millet** (Nagli-RAU-8)
$FN = 4.42\ T - 0.225\ SN$
$FP_2O_5 = 2.97\ T - 1.32\ SP$
$FK_2O = 1.21\ T - 0.024\ SK$

m) **Fodder Maize** (African tall)
$FN = 6.49\ T - 0.56\ SN$
$FP_2O_5 = 1.51\ T - 0.73\ SP$
$FK_2O = 2.45\ T - 0.13\ SK$

n) **Soybean** (JS-335)
$FN = 6.86\ T - 0.68\ SN$
$FP_2O_5 = 6.17\ T - 4.46\ SP$
$FK_2O = 3.96\ T - 0.13\ SK$

o) **Sunflower** (SS-56)
$FN = 13.94\ T - 0.61\ SN$
$FP_2O_5 = 7.18\ T - 6.82\ SP$
$FK_2O = 4.82\ T - 0.12\ SK$

p) **Pigeon pea** (ICPL-87)
$FN = 5.61\ T - 0.54\ SN$
$FP_2O_5 = 5.72\ T - 4.73\ SP$
$FK_2O = 6.33\ T - 0.17\ SK$

q) **Green gram** (S-8)
$FN = 4.56\ T - 0.18\ SN$
$FP_2O_5 = 12.51\ T - 7.61\ SP$
$FK_2O = 3.53\ T - 0.05\ SK$

r) **Chick pea** (Vishal)
$FN = 5.25\ T - 0.46\ SN$
$FP_2O_5 = 3.87\ T - 2.77\ SP$
$FK_2O = 1.29\ T - 0.04\ SK$

s) **Onion** (N-2-4-1)
$FN = 5.40\ T - 0.54\ SN$
$FP_2O_5 = 4.00\ T - 4.32\ SP$
$FK_2O = 3.10\ T - 0.13\ SK$

Chapter 10

Spatial Data and their Management in GIS

CHAPTER OUTLINES

10.1 Definitions

Spatial data is the geographic position of a particular geographic feature using coordinate systems. *e.g.* **spatial data** can be seen in a road map. A road map is a two-dimensional object that contains points, lines and polygons that can represent cities, roads and political boundaries such as states or provinces. A road map is a visualization of geographic information. The most common method of **collecting spatial data** is the use of **global positioning system** (GPS) receivers.

Spatial data is the data which include geographic location information of a point, line or polygon object like shape, size and location of the feature. **Non-spatial data** refers to other attributes associated with the feature such as name, length, area, volume, population, soil type or **Attribute data** is the data which include any other non-location information related to a point, a line or a polygon. Attribute data comprise the pertinent information about the spatial data. The querying feature works based on attribute data.

10.2 Importance of Spatial Data

a) To solve complex location-oriented problems
b) Better understand where and what is occurring in the world
c) To study the characteristics of places
d) To study the relationships between places
e) Spatial analysis lends new perspectives to your decision-making

10.3 Representation of Spatial Data

a) **Geometry:** Geometry data are represented as points on a planar or flat-earth surface. *e.g.* (5,2) where the first number represents that point's position on the horizontal (x) axis and the second number represents the point's position on the vertical (y) axis.

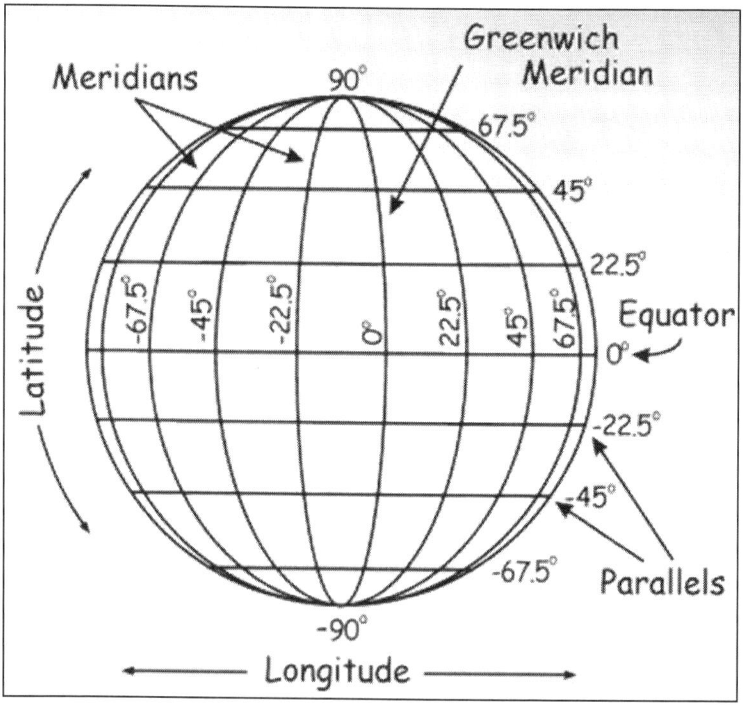

Figure 10.1. Latitude and Longitude

b) **Geography:** Geography data are represented as latitudinal and longitudinal degrees as on earth or other earth-like surfaces. An example would be **19.58°N latitude** and **74.20°E longitude** showing Sangamner area of Ahmednagar district.

c) **Geographic coordinate system:** To identify exact locations on the surface of the earth, a geographic coordinate system is used. Normally, an x and y-axis are used, the axes are referred to as lines of latitude (horizontal lines) and longitude (vertical lines). Each axis represents the angle at which that line is oriented with respect to the center of the earth, hence, the units are measured is degrees (^{o}).

d) **Georeferencing** is the process of assigning coordinates to vectors or rasters so they can be oriented accurately on a model of the earth's surface.

e) **Geocoding** is the process of converting addresses into geographic coordinates which use to place markers on a map or position the map.

Geoprocessing is a framework and set of tools for processing geographic and related data. Geoprocessing can be used to perform spatial analysis or manage GIS data in an automated way.

10.4. Application of Spatial Data

a) GIS (Geographic Information System): *e.g.* Map, graphic.

b) MMIS (Multimedia Information System): *e.g.* Video, audio, image, text.

c) CAD (Computer Aided Design): *e.g.* Surface of design object.

10.5 Types of Spatial Data in GIS

There are two types of spatial **data** used in **GIS**, namely: Vector data and Raster data. Each representation has merits and is suited for particular **kinds** of information and analysis;

10.5.1 Vector Data

A vector feature has its shape represented using geometry. The geometry is made up of one or more interconnected vertices. The vertex describe a position in space using an X, Y and optionally Z axis and represented as points, lines, polygons and triangulated irregular networks. Vector data is best described as graphical representation of the real world. There are three main types of vector data: points, lines, and polygons. Connecting points create lines, and connecting lines that create an enclosed area create polygons. Vectors are best used to present generalizations of objects or features on the earth's surface.

a) Point Data

i) Represented as a single 'dot 'on the map

ii) Points are used to indicate discrete locations

iii) They usually have a single X, Y coordinate

iv) Used to represent a feature that is too small to be displayed as a line or area

b) Line Data

i) Ordered sets of points that have the look of a straight line or a curved arc depending upon the feature it describes

ii) They have a length but no width

iii) Used to represent a geographical feature such as a stream or a road.

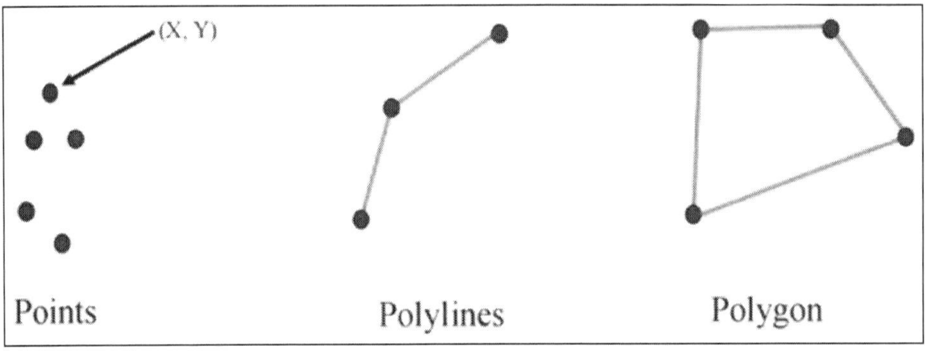

c) Polygon Data

i) Closed features whose boundary encloses a homogenous area

ii) Have an area that is given by the arcs/lines that make the boundary

iii) Used to represent features that have area

Advantages

a) Accurately representing true shape and size

b) Representing non-continuous data (River, road live, political boundaries)

c) Store information about topology

d) Data are easily overlaid, *e.g.* roads, rivers, land use are easier than raster data

e) Data are more compatible with the relational database management system.

f) Vector file sizes are much smaller than raster image file

g) Vector data are easier to update like adding river stream

Disadvantages

a) Location of each vertex needs to be stored explicitly

b) Requires extensive data cleaning

c) Updating or editing of the vector data requires re-building of the topology

d) Data must be converted into a topological structure

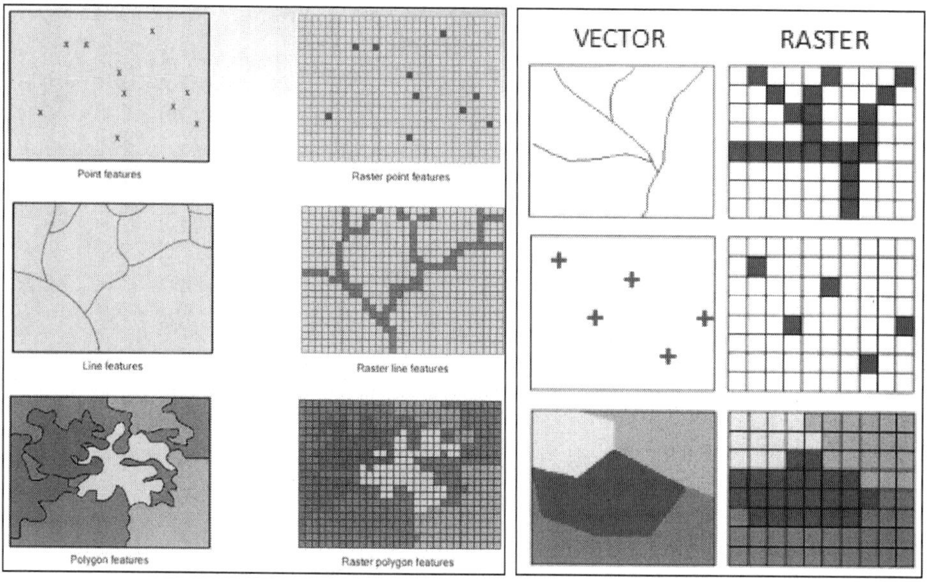

Figure 10.2. Raster Data and Vector Data in GIS.

10.5.2 Raster Data

Raster data are composed of grid cells identified by row and column. The whole geographic are divided into groups of individual cells which represent an image. *e.g.* Satellite images, photogrammetry and scanned. Raster models are used to store data, which varies continuously as in aerial photography, a satellite image or

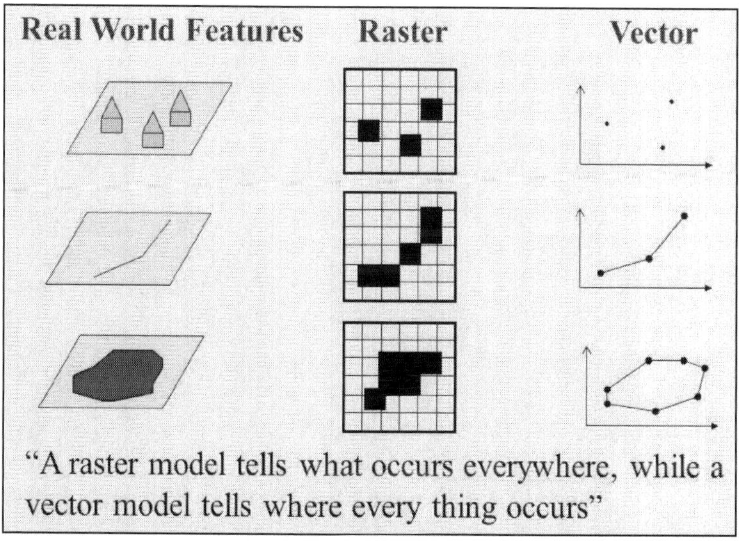

Figure 10.2. Comparison between Vector Data and Raster Data.

elevation values. Raster data is characterized by pixel values, each pixel is assigned a specific value from 0 to 255. A raster data set is a regular grid of cells divided into rows and columns. In a raster data set, data values for a given parameter are stored in each cell. These values may represent an elevation in meters above sea level, a land use class. The spatial resolution of the raster data set is determined by the size of the cell, these cells are used to represent geographic data.

Advantages

a) Raster data model record value of all the points of the area covered which required more data storage than model represented by the vector model
b) Store continuously changing values such as elevation, slope
c) Raster data is less expensive to create computationally compare to vector graphics
d) Analysis faster and more flexible then vector for many application

Disadvantages

a) Difficult to adequately represent linear features depending on the cell resolution
b) Processing of associated attribute data may be cumbersome if large amounts of data exists
c) Raster maps inherently reflect only one attribute or characteristic for an area
d) Most output maps from grid cell systems do not conform to high quality cartographic needs.

Chapter 11

Image Processing and Interpretation

CHAPTER OUTLINES

11.1 Image Processing

Image processing is a method to convert an image into digital form and perform some operation on it, in order to get an enhanced image or to extract some useful information from it.

11.1.1 Purpose of Image Processing

a) Visualization: observe the objects that are not visible
b) Image sharpening and restoration: to create a better image
c) Image retrieval: seek for the image of interest
d) Measurement of pattern measure various objects in an image
e) Image recognition: Distinguish the object in an image

11.1.2 Types of Image Processing

a) **Analog or visual image processing:** It is any image processing task conducted on two dimensional analog signals by analog means. If the pictorial representation of the data represented in analog wave formats that can named as analogue image. Analog image are required for human viewing. It contains various levels of brightness and color. It is generally continuous and not broken into many small individual pieces.

b) **Digital image processing:** The digital representation or storing the data in digital form is termed as digital image processing. Digital image processing is the use of a digital computer to process digital images through an algorithm.

Advantages of Digital Image Processing

a) Cost effective for large geographic areas
b) Cost effective for repetitive interpretations
c) Cost effective for standard image formats
d) Consistent results
e) Simultaneous interpretations of several channels
f) Complex interpretation algorithms possible
g) Speed may be an advantage
h) Compatible with other digital data

Disadvantages of Digital Image Processing

a) Expensive for small areas
b) Expensive for one-time interpretations
c) Start-up costs may be high
d) Requires elaborate, single-purpose equipment
e) Accuracy may be difficult to evaluate
f) Requires standard image formats
g) Data may be expensive or not available
h) Preprocessing may be required
i) May require large support staff

11.2 Procedures Used in Analyzing Remote Sensing Images

a) Pre-processing
b) Image enhancement
c) Image classification
d) Spatial feature extraction
e) Measurement of bio-geophysical parameters
f) GIS

a) Pre-processing

Raw data is usually carried out to corrected for any distortion due to the characteristics of the imaging system and condition. Some standard correction procedures may be carried out by the ground station operator before the data is delivered to the end-user.

i) **Radiometric correction:** to correct for uneven sensor response over the whole image.

ii) **Geometric correction:** to correct for geometric distortion due to earth's rotation and other imaging conditions. The image may also be transformed to conform to a specific map projection system. Accurate geographical location of an area on the image needs to be known ground control points are used to register the image to a precise map.

b) Image enhancement

Image enhancement is a digital technique to improve the appearance of an image for human visual analysis and machine analysis. Image enhancement improved by visual appearance of the objects in the image appearance of the objects in the image.

i) **Grey level stretching:** to improve the contrast and spatial filtering for enhancing the edges. The results of applying the grey level stretching is that the hazy appearance due to scattering of sunlight by atmosphere into the field will be removed.

ii) **Linear grey level stretching:** A level threshold value is chosen so that all pixel values below this threshold are mapped to zero and also upper threshold value is chosen so that all pixel value above this threshold are mapped to 255. All other pixel values are linearly interpolated to lie between 0 and 255. The result of applying the linear stretch is that the hazy appearances will be removed except for some parts near to the top of the image.

c) Image Classification

There are two methods of digital image classification;

i) **Supervised classification:** There are six stages of the classification process; a) To define training sites, b) Extract signatures, c) Classify the image, d) In-process classification assessment, e) Generalization and f) Accuracy assessment. select training samples and classify your image based on your chosen sample. **Training area-** spectral features of some areas of known land covers types are extracted from an image.

ii **Unsupervised classification:** The computer programmer automatically groups the pixels in the image into separate clusters, depending on their spectral features, each cluster will then be assigned a land covers type by the analyst. **Theme -** each class of land covers. **Thematic map-** Product of classification.

d) Spatial Feature Extraction

Pixel-based method can be used in the lower resolution mode and merged with the contextual and textural method at higher resolution. To fully exploit the spatial information contained in the imagery, image processing and analysis algorithms utilizing the textural contextual and geometrical properties are required. Algorithm make use of the relationship between neighboring pixels for information extraction.

e) Measurement of Bio-geophysical Parameters

Specific instruments carried on board the satellites to continuously monitor the global variation of environmental factors (sea, water, chlorophyll concentration, Atmosphere, vapour pressure, water vapour, ozone, land and sea temperature, forest biomass, aerosol and sea surface wind field.

f) Geographical Information System

Optical and radar images provides complementary information about the land cover. More detailed information can be derived by combining several different types of images. The thematic information derived from the remote sensing image are often combined with other auxiliary data to form the basis for a GIS. GIS is a database of different layers, where each layer contain information about a specific aspect of the same area.

11.3 Image Interpretation

Interpretation is the process of detection, identification, description and assessment of significant of an object and pattern imaged.The ERDAS IMAGINE, IDRISI, ILWIS, ENVI, eCognition are some software brands, which facilitate digital image interpretation.

11.3.1 Types of Image interpretation

A) Visual Interpretation

Aerial photographs and satellite imageries are interpreted visually.

a) **Aerial photograph:** The aerial photography is defined as the science of making photographs from air craft's for studying the earth surface. Photogrammetry is the science which study interpretation of aerial photographs. Pocket stereoscope, mirror stereoscope, plotter is used in photogrammery for measuring area, height, slopes of different parts of earth. photographed and also for plotting different objects/theme from aerial photographs.

b) **Satellite image interpretation**: is an art of examining images for the purpose of identifying objects and judging their significance

B) Digital Interpretation

It is the analysis of digital data with the help of computer to extract information about the earth surface Digital interpretation is populary known as Image processing.

11.3.2 Basic Elements of Interpretation

Interpretation of satellite imagery and aerial photographs involves the study of various basic charter of an object with reference to spectral bands which is useful in visual analysis.

a) **Shape** - The external form, outline or configuration of the object. this includes natural features *e.g.* Yamuna River in Delhi man made feature.

b) **Size** - Size is the scale of an image. This property depends on the scale and resolution of the image/photo smaller feature will be easily indented in large scale image/photo.

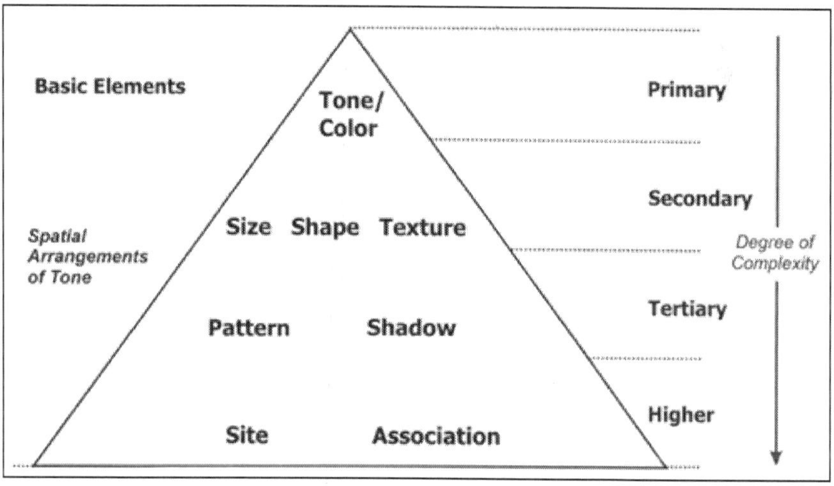

Figure 11.2. Ordering of Image Elements in Image Interpretation.

c) **Pattern** - Pattern is the spatial arrangement of an object into distinctive recurring forms. This can be easily explained through the pattern of a road and railway line, even though both looks linear major roads associated with steep curve and many intersection with minor road.

d) **Shadow** - It indicates the outline of an object and its length which is useful is measuring the height of an object the shadow effect in radar images is due to look angle and slope of the terrain. taller feature cast large shadows than shorter features.

e) **Tone/Colour** - Refers to the color or relative brightness of an object. The tonal variation is due the reflection, emittance, transmission or absorption character of an objects. This may vary from one object to another and also changes with reference to different bands.

f) **Texture -** It is the arrangement and frequency of colour changes in particular areas of an image. It creaks a visual impression of surface roughness or smoothness of objects.

g) **Site -** Site is the topographic or geographic location. It is important in the identification of vegetation types.

h) **Resolution -** It depends upon the photographic device namely cameras or sensors. This includes of spectral and spatial resolutions. The spectral resolution helps in identifying the feature in specific spectral bands. The spatial resolution imagery is useful in identifying small objects.

i) **Association -** It is the relationship between the objects and their location.

	What is it?	**Example**
Tone	Refers to the relative brightness or colour of objects in an image	Can distinguish between two crop fields due to different texture and shape of plants
Shape	Refers to the general form, structure, or outline of individual objects	Manmade objects tend to have straight edges, whereas natural objects have more irregular shapes
Size	**Size** of objects in an image is a function of scale	It is easy on an image to distinguish between a football stadium and a house
Pattern	Refers to the spatial arrangement of visibly discernible objects	In a housing estate, although individual houses cannot be made out, the recognisable pattern its produced
Texture	Refers to the arrangement and frequency of tonal variation in particular areas of an image	Different textures will appear differently on an image, for example, a sandy clearing within a forest will show up on the image
Shadow	May provide an idea of the profile and relative height of a target or targets which may make identification easier.	Shadow allows for identification of topographical landforms
Association	**Association** takes into account the relationship between other recognizable objects or features in proximity to the target of interest	Boats cannot be made out on their own within an image, however when in a marina, their proximity allows for identification

11.3.3 Issues in Interpretation

a) Unfamiliar scale and resolutions

b) Lack of understanding of physics of remote sensing

c) Understanding proper spectral character of each object

d) Visually interpret three layers of information at a time

Chapter 12

Nanotechnology

CHAPTER OUTLINES

12.1 Definitions

Nanoscience is the study of phenomena and manipulation of materials at atomic, molecular and macromolecular scales. **Nanotechnology** is the application of nanoscience to produce devices and products at dimensions between 1 to 100 nanometers(nm). **or Nanotechnology** is the design, characterization, production and application of structures, devices and system by controlling shape and size at nanometer.

12.2 Concepts of Nanotechnology

The word "nano" comes from **Gree**k means "dwarf" or small. A nanometer (nm) is a thousandth of a meter (10^{-9} m). The ideas of nanotechnology was first time introduced in **1959** by **Richard Feynman** at Caltech. He never mentioned, "nanotechnology" but he suggested that it will eventually be possible to precisely manipulate atoms and molecules. The "nanotechnology" was first used in **1974** by **Noria Taniguchi** and refer to the ability to engineer materials precisely at the scale of nanometers.

12.2.1 Macro Scale, Micro Scale and Nano Scale

a) **Macro:** *e.g.* i) Person height is about 180 cm *i.e.*, approx. 2 billion nm; ii) Apple size is about 8cm *i.e* 80 million nm, iii) Ant is about 5 cm *i.e* 5 million nm

b) **Micro:** *e.g.* i) human hair is about 80,000nm diameter, ii) *E. coli* bacteria is about 2000nm.

c) **Nano:** *e.g.* i) DNA- diameter 2nm, ii) Hydrogen atom- 0.1 nm

12.2.2 Nanomaterials

Nanoparticles is the small object that act a whole unit in terms of transport and properties.

Properties of Nanomaterials

a) Large surface area
b) Materials more chemically reactive
c) Quantum effect can begin to dominate the behavior of matter at the nanoscale
d) Highly mobile in the free state
e) Exhibit different colour depending upon their size
f) Reactivity of nanomaterials is more than that of bulk materials
g) Nanoscale should not constant physical properties regarless of its size
h) Size dependent properties are observed such as quantum confinement confinement in semiconductor particles and super paramagnetic in magnetic material

Advantages of Nanomaterials

a) Surface area increases as decreases
b) Perform very faster
c) Lighter than bulk materials
d) Can get into small spaces
e) Cheaper than bulk materials
f) More energy efficient
g) Show different properties at very small scale

12.2.3 Two Main Approaches are Used in Nanotechnology/Fabrication Techniques

a) **Top down approach**: Starting with a larger components and carving away materials. *e.g.* Rock to clay soil

b) **Botton up approach:** Assembling smaller components, atoms by atom assembly.

e.g. Bricks to building.

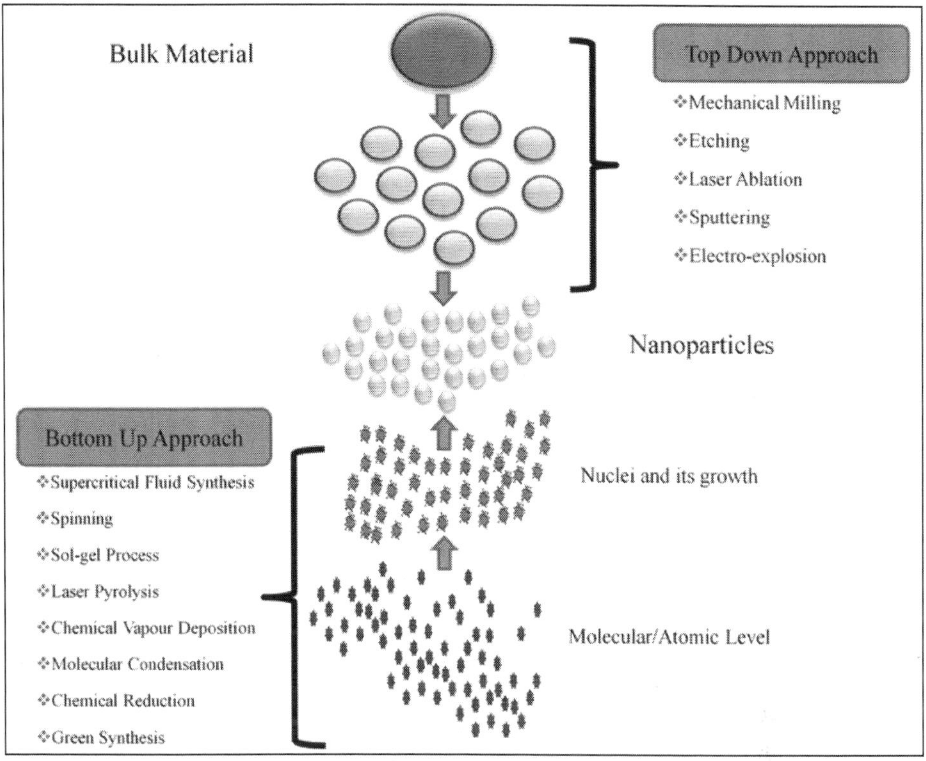

Figure 12.1. Synthesis of Nanomaterials (NPs) via Top-down and Bottom-up Approaches.

12.2.4 Nanoscale

Nanoscale is the structures, design with a length scale application to nanotechnology usually 1nm to 100 nanometers.

12.2.4.1 Category of Nanomaterials According to Dimensions

a) **One dimensions:** it has only one parameters (length/breath/height), 1 dimension = <100 nm. *e.g.* Thin film, layers, coating

b) **Two dimensions:** it has only length and breath, 2 dimension = <100 nm length and breath. *e.g.* nano-tubes, nano- fibres, nano-wires

c) **Three dimensions:** it has all parameters (length, breath, height), 3 dimension = < 100 nm. *e.g.* nanoparticles, Nanoshell, Nanorings.

12.2.4.2 Surface Area to Volume Ratio

Cube	*Formula*	*2mm x 2mm x 2mm Cube (1)*	*Break Cube in 1mm x 1mm x 1mm*
Surface area	Height x Width x No. sides x No. of cubes	2 x 2 x 6 x 1 = 4	1 x 1 x 6 x 8 = 48
Volume	Height x Width x Length x No. of cubes	2 x 2 x 2 x 1 = 8	1 x 1 x 1 x 8 = 8
Surface volume ratio	Surface area/Volume	24/8 = 3	48/8 = 6

If surface to volume ratio increases then properties we achieve this are,

a) A greater amount of a substances comes in contact with surrounding materials

b) Hidden surface are exposed

c) Greater proportional of the materials is exposed for potential reaction

12.2.4.3 Nano-scale Size Effect

a) Physical properties: melting point. Lower the size lower the melting point.

b) Chemical properties: reactivity. Reactivity increases with nanoscale

c) Electrical properties: Non-conductor material show conductor property at nanoscale.

d) Mechanical properties: Strenght (high strength and efficiency)

e) Optical properties: light emission *e.g.* gold (sphere) at 50nm show green colour, Silver (sphere) at 40 nm show Blue colour, Gold (sphere) at 100 nm show Orange colour.

12.3 Tools and Techniques

12.3.1 Characterisation of Nanomaterials

a) UV- visible spectroscopy

b) Dynamic light scattering

c) Zeta potential

d) Electron microscopy

e) Atomic force microscopy

12.3.1.1 UV-visible Spectroscopy

It is first steps in characterization of nanoparticles. *e.g.* Dispersed gold nanoparticles is in red colour and is aggregated it is in purple colour.

12.3.1.2 Zeta Potential

The stability of nano-particles can be studied by zeta potential. Zeta potential is the overall charge that the particle acquire in a particular medium. Zeta potential values should be = or > +/- 30 milli volt indicates good stability. It is depends on pH and electrolyte concentration. *e.g.* Zeta sized equipment.

12.3.1.3 Dynamic Light Scattering (DLS)

It is for measurement of particle size. Size will include any stabilizer bounds to the molecules scattering of laser light due to Brownian motion of particles. It measures hydrodynamic diameter of particles at range 0.6nm – 6 micron

12.3.1.4 Microscopy

Microscopy is any techniques for producing visible images of structure.

1931 Ernst Ruska invents the electron microscope

1938 10 nm resolution reached

1940 2-4 nm resolution reached

1945 1.0 nm resolution reached

1981 Gerd Binning and Heinrich Rochrer (Noble in 1984) invent the scanning tunneling electron microscope (STM)

1986 Gerd Binning and Heinrich Rochrer invent the Atomic force microscope

a) Transmission Electron Microscope (TEM)

TEM is a complex viewing system equipped with a set of electromagnetic lenses used to control the imaging electrons in order to generate the extremely fine structural details that are usually recorded on photographic film. It gives information about the size of NPS morphology, crystallographic information and compositional information.

b) Scanning Electron Microscope (SEM)

SEM is a microscope that produces an image by using an electron beam that scans the surface of a specimen inside a vaccum chamber. It gives the information about topograph morophology, elemental composition, crystallography and orientation.

c) Scanning Probe Microscopes (SPM)

It monitors the interactions between a probe and sample surface.

Types of Scanning Probe Microscopes

a) Scanning tunneling microscope (STM)

b) Atomic force microscope (AFM)

c) Scanning confocal microscope (Marvin Minsky-1961)

d) Scanning acoustic microscope (Calvin Quate-1970)

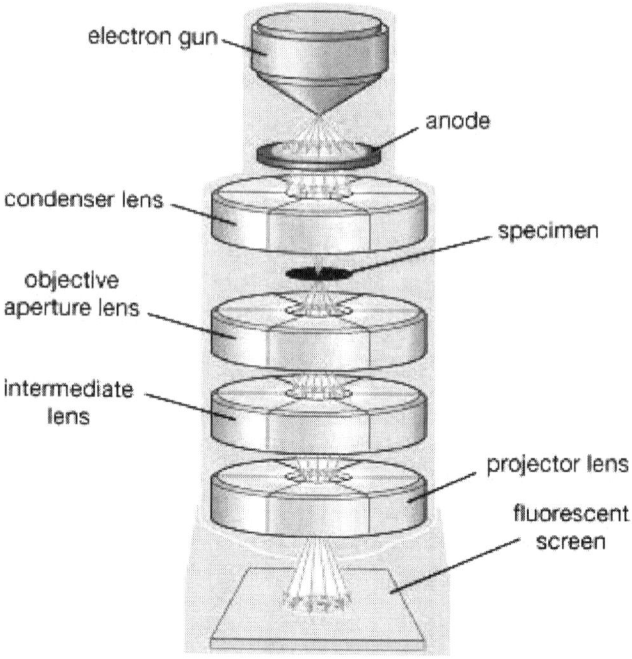

Figure 12.2. Transmission Electron Microscope.

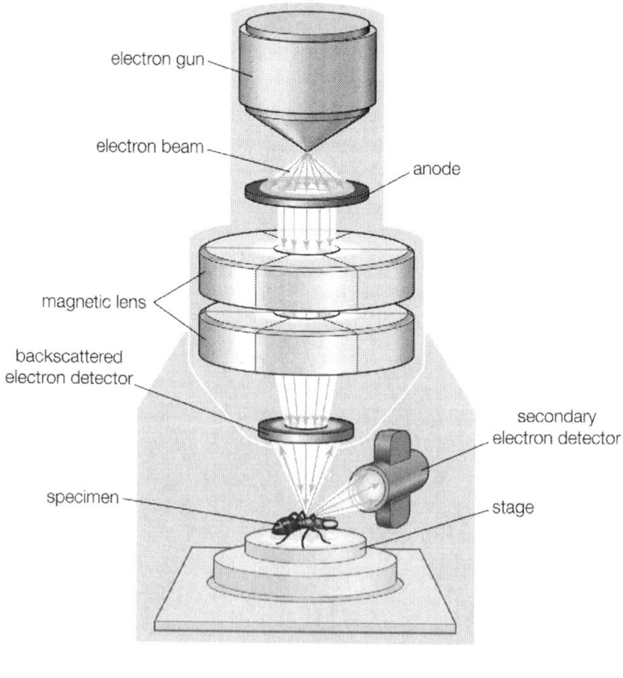

Figure 12.3. Scanning Electron Microscope.

i) Scanning Tunneling Microscope

STM monitors the electron tunneling current between a probe and a sample surface tunneling current depends on the distances between the STM probe and the sample and gives as atomic information about the surface.

ii) Atomic Force Microscope (AFM)

AFM monitors the forces of attraction and repulsion between a probe and sample surface. The tip is attached to a cantilever which moves up and down in response to forces of Attraction of repulsion with the sample surface and movement of the cantilever is detected by a laser and photodector.

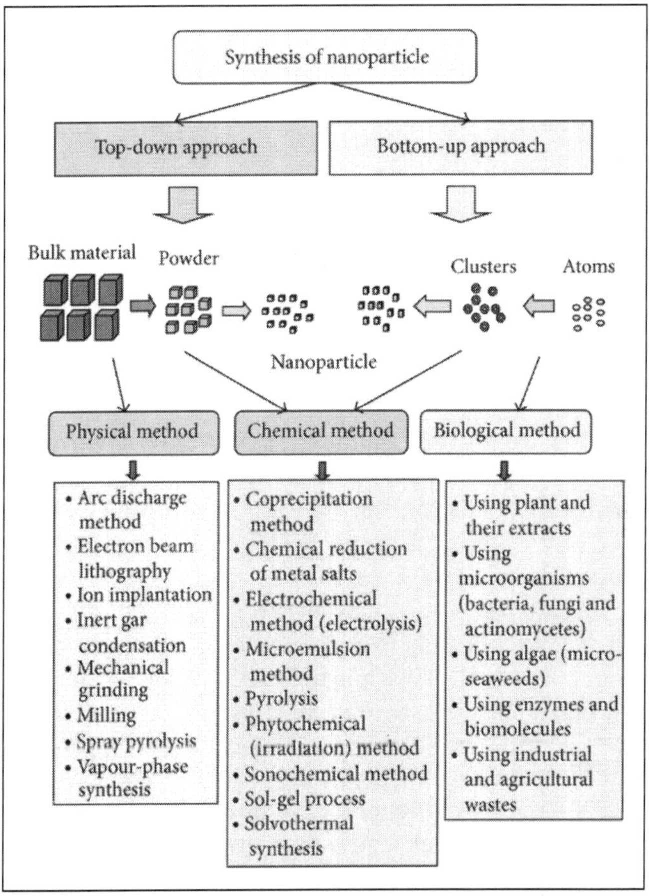

Figure 12.3. Different Approaches and Methods for Synthesizing Nanoparticles.

Application of AFM

a) Study topographical surface

b) Determine the roughness of a surface sample

c) Measure the thickness of a crystal growth layer
d) Study of dynamic behavior of living and fixed cells
e) Not required vaccum
f) High resolution 0.1 nm or 1nm X –Y direction

12.3.2 Synthesis of Nanomaterials

A) Physical method
B) Chemical method
C) Biological method

A) Physical Method

i) High energy ball milling
ii) Laser ablation
iii) Electron beam lithography
iv) Inert gas condensation
v) Physical vapour deposition
vi) Spray pyrolysis
vii) Flash spray pyrolysis

i) High Energy Ball Milling (HEBM)

It is first developed by John Benjamin in 1970. It is top down method. It is a ball milling process where a powder mixture placed in the ball mill is subjected to high energy collision from the balls, the moving balls transfer their kinetic energy to the milled material. This results in the breaking of their chemical bonds and rupturing of the milled materials into smaller particles with newly created surfaces. Milling media, milling speed, ball-to-powder weight ratio, type of milling, type of high energy ball mill, milling atmosphere and duration of milling regulate the amount of energy transfer between the balls and the material during the process, and thus affect the physical and morphological properties of the resultant nanomaterials.

ii) Laser Ablation

In the laser ablation technique, a high power laser was used to vaporize carbon from a graphite target at high temperature. **LASAR (**Light Amplification by Stimulated Emission of Radiation**).** It involves the generation of nanoparticles by laser ablating a solid target that lies in a gaseous or a liquid environment and collection of the nanoparticles in the form of nanopowder or colloidal solutions. Both multi walled nanotube and single walled nanotube can be produced with this technique.

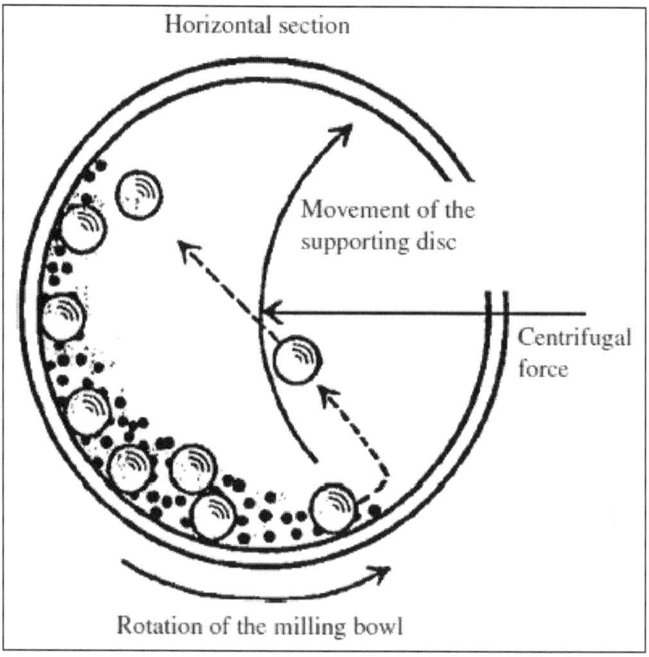

Figure 12.4. High Energy Ball Milling Method.

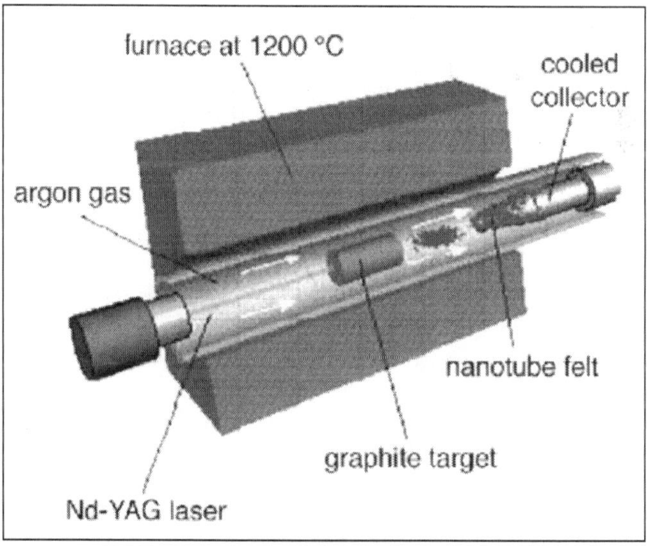

Figure 12.5. Schematic Diagram of the Laser Ablation Apparatus.

Advantages of Laser Ablation Method

a) It is an easy, fast and straight forward method of NPs synthesis than forward method of NPs synthesis than other

b) It does not require long reaction times or multi-step chemical synthetic procdures
c) It does'n require the use of toxic/hazardous chemical precursors for synthesis of NPs
d) Laboratory safe method
e) Known as environmentally friendly

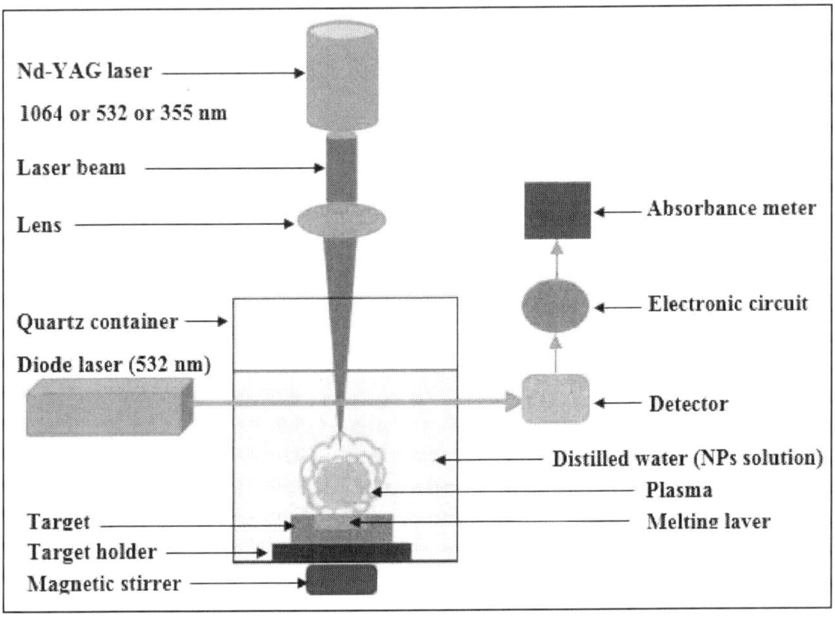

Figure 12.6. Synthesis of Nanoparticles by Laser Ablation.

B) Chemical Method

i) Sol-gel method
ii) Micro emulsion technique
iii) Hydrothermal synthesis
iv) Polyol synthesis
v) Chemical vapour synthesis
vi) Microwave assisted synthesis
vii) Plasma enhanced chemical vapour deposition technique

The components needed for chemical methods; a) Metal salts, b) Reducing agent and c) Capping agent. **Reducing agent**: reduce metal salts into metal nano-particles. *e.g.* sodium nitrate, $NaBH_4$, Ascorbic acid. **Capping agents**: nano- metal particles are high unstable and try to join together and form aggregate for making nano-particle stable, capping agents should apply that prevent the aggregation of nanoparticles. *e.g.* Thiols, citrate, PVA, PVP (Polymers).

For Synthesis of Nanoparticles Involves

1) Nucleation and
2) Growth

a) Nucleation

Nucleation is the phenomenon of initiation of formation of the first nano-crystals in solutions. It involves the appearance of very small particles or nuclei of the new phase which are capable of growing.

Types of Nucleation

1. Homogenuos nuclei: form simultaneously and uniformly throughout the solution
2. Heterogenous nuclei: Nucleated at different time

Homogeneous Nucleation: All nuclei will be form simultaneously and uniformly will get monodispersed particles, uniformly size particles. **Heterogonous nucleation:** nucleated at different time will get polydispersed particles. It play important role in controlling the properties of the final product, size distribution and nature of the phase.

For Nanoparticles need strong nucleation and slow growth. If growth continue, the particles structure so big hence we need to stop growth because we need small size particles for application

Role of Stabilizing Agents/Caping Agent/Ligands

a) Prevent uncontrollable growth of particles
b) Prevent particle aggregation
c) Control growth rate
d) Control particle size
e) Allows particle solubility in various solvents

Stabilization of Nano Clusters against Aggregation

a) **Electrostatic stabilization**: Absorption of ions to the surfaces. It creates on electrical double layer which results in a coulombic repulsion force between individual particles.
b) **Steric stabilization**: metal centers surrounding by layer of sterically bulky materials. *e.g.* polymer, surfactants.

Parameters Affecting Particle Growth

a) Type of capping agent
b) Reducing agent
c) Concentration of the reactants
d) pH value of the solution
e) Duration of heat treatment

I) Sol Gel Method

The sol gel process is a method for producing solid materials from small molecules. The method is used for the fabrication of metal oxides, especially the oxides of silicon (Si) and titanium (Ti). The initial step of this process is converting monomers or the starting material into a sol, *i.e.*, a colloidal solution which is the precursor for the further formation of a gel. This gel is made up of discrete particles

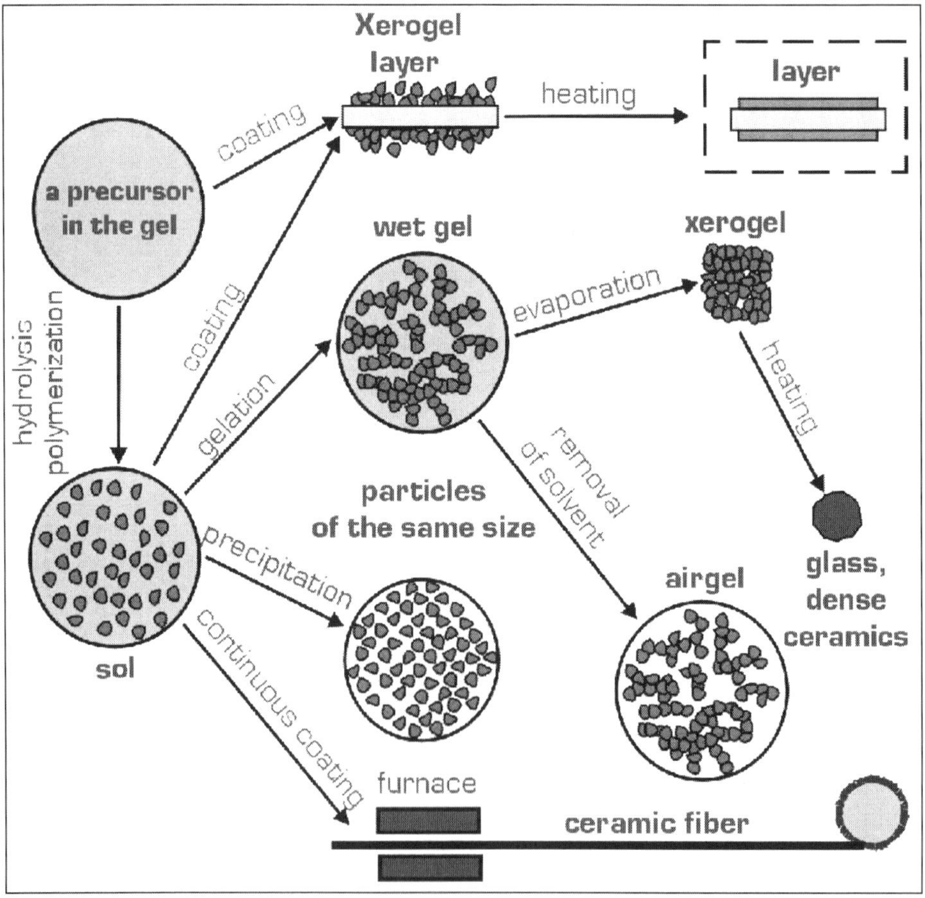

Figure 12.7. Schematic Representation of the different Stages and Routes of the Sol Gel Technology.

or polymers. Most commonly used precursors are chlorides or metal alkoxides. These precursors are hydrolysed and poly condensed for the formation of colloids. Sol or colloidal solution is a solution where distribution of particles of the size ~ 0.1-1μm takes place in a liquid in which the only suspending force is the Brownian motion. A gel is formed when solid and liquid phases are dispersed in each other. In this process, initially, colloidal particles are dispersed in a liquid forming a sol. Deposition of this sol can produce thin coating on any substrate by the means of spraying, spinning or coating. The particles in the sol are left to polymerize by removing the stabilizing components and further produce a complex network gel. The remaining organic and inorganic components pyrolyze in the end by heat treatments to form amorphous or crystalline coatings. Precursor sol which is obtained can be given a desired shape using appropriate casting container. It can also be deposited on a substrate to form a film by dip coating or spin coating or used to synthesize microsphere or nanosphere powders. Sol-gel process is preferred due to its economical feasibility and the low-temperature process which gives us control over the composition of the product achieved.

II) Hydrothermal Synthesis

This method is used to fabricate NPs of metal oxide, iron oxide and lithium iron phosphate keeping control over the characteristics of particles by varying the properties of near or supercritical water by using different pressure and temperature conditions. The advantage of using this method includes the capability to synthesize a huge amount of NPs with an optimized size, morphology, composition and surface chemistry that is rationally inexpensive. Hydrothermal is a facile and fast process for the synthesis of NPs of various other materials such as $CoFe_2O_4$, Ag, $FeWO_4$, CdS, Zr, ZnO, *etc.*

C) Biological Methods

i) Use of microorganisms
ii) Use of plant extract/enzymes
iii) Use of DNA, virus

Advantages of Biological Methods

a) Reduce toxic chemicals concentration
b) Eco-friendly nano-particles
c) Economically viable
d) Easier to formation

Green method similar to chemical method except chemical is replace by microbial enzymes or plant photochemical. Yeast, bacteria, fungi, leaves, stem, roots, flower *etc.* used as reducing and capping agent in biological method.

Synthesis of Metallic Nanoparticles by different Microorganisms

Microorganism	*Types of nanoparticles*	*Location*	*Size (nm)*
a) Bacteria			
Pseudomonas stutzeri	Ag	Intracelluar	200
Lactobacillus strains	Ag and Au	Intracelluar	
Cyanobacteria	Ag	Intracelluar	1-10
Morganella spp.	Ag	Extracelluar	
Actinobacter spp.	Magnetite	Extracelluar	10-40
b) Yeast			
Candida glabrata	Cds	Intracelluar	200
Schizosaccharomyces pombe	Cds	Intracelluar	200
c) Fungi			
Fusarium oxysporum	Au	Extracelluar	20-40
Trichoderma asperellum	Ag	Extracelluar	13-18
Aspergillus Fumigutus	Ag	Extracelluar	5-25
Verticillium	Ag	Extracelluar	12-25
d) Algae			
Chlorella Vulgaris	Au	Extracelluar	9-20
e) Plant			
Azadirachta indica	Au/Ag	Extracelluar	50-100
Geranium leaves	Ag	Extracelluar	16-40

The following important parameters play a significant role in the biosynthesis of nanoparticles.

a) **Bioresources used for nanoparticle biosynthesis:** The synthesis of nanoparticles is characterized by choice of the most convenient microorganism with respect to: growth rate, enzyme production and the respective metabolic pathways. Some microorganisms, like bacteria, viruses, fungi, yeasts and algae, are used for the biosynthesis of metallic nanoparticles.

b) **Cellular metabolites involved in biosynthesis:** Molecules like enzymes, proteins, polysaccharides *etc.* act as reducing and stabilizing agents in the biosynthesis of nanoparticles. They can be utilized in the process as whole-cell microorganisms, crude cell preparations and crude or purified enzymes obtained from microorganisms.

c) **Reactions facilitating nanoparticle biosynthesis**: The process of this biosynthesis is initiated with harvesting of microbial biomass, which is related with residual nutrients and metabolites to avoid undesired by-product reactions. During the processes of scalling up, the production rate and product yield are of special interest and optimization is necessary (*e.g.* production time, pH, temperature *etc.*). The process of optimization of these factors can influence the particles morphology and their properties.

d) **Growth of inoculum for biosynthesis of nanoparticles:** The biosynthesis of nanoparticles depends on the growth conditions of the microbial producers, *e.g.* nutrients, pH, temperature, *etc.* These factors need to be optimized.

e) **Microbial nanoformulations:** exploring the potential for nano-farming. Nanoparticles synthesized by microbes are highly stable and could offer a non-toxic, cost-effective and eco-friendly approach for synthesis over chemical ones. This green synthesis has a great advantage over the chemical methods, which have toxic effects on the environment. Thus, the use of agriculturally important microorganisms for nanoparticle biosynthesis and their further role in agriculture is of considerable significance. The use of nanoformulations may enhance the stability of bio-fertilizers and bio-stimulators with respect to desiccation, heat, and UV inactivation.

12.4 Application of Nanotechnology in Agriculture

a. Soil Improvement

i) Zeolite and nanoclays for agrochemicals, water retention and their slow release in soil

b. Crop Production

i) Nanofertilizers for enhancing nutrient absorption by plant, to increase nutrient efficiency and improve plant nutrition compared with traditional fertilizers

ii) Nanofungicides for enhancing the management of fungal diseases of food and fruits crops

iii) Nonopesticide for enhancing the efficacy or reduce the environmental footprint of a pesticide active ingredient

iv) nanoherbicides for reducing the herbicide load on the environment also help in eradication of weeds without leaving any major toxic residues in soils and environment

v) Nanoencapsulation for reducing the current application dosage and improving efficiency of a fertilizer or pesticide

vi) Nanoparticles used in the preparation of new formulations of fertilizers and pesticides

vii) Nanoemulsions for increase uptake and effectiveness

c. Precision Farming

i) Nanosensors for plant pathogen and pest detention

ii) Nonosensors for monitoring soil conditions

d. Crop Improvement

i) Nonomaterial assisted plant genetics modifications

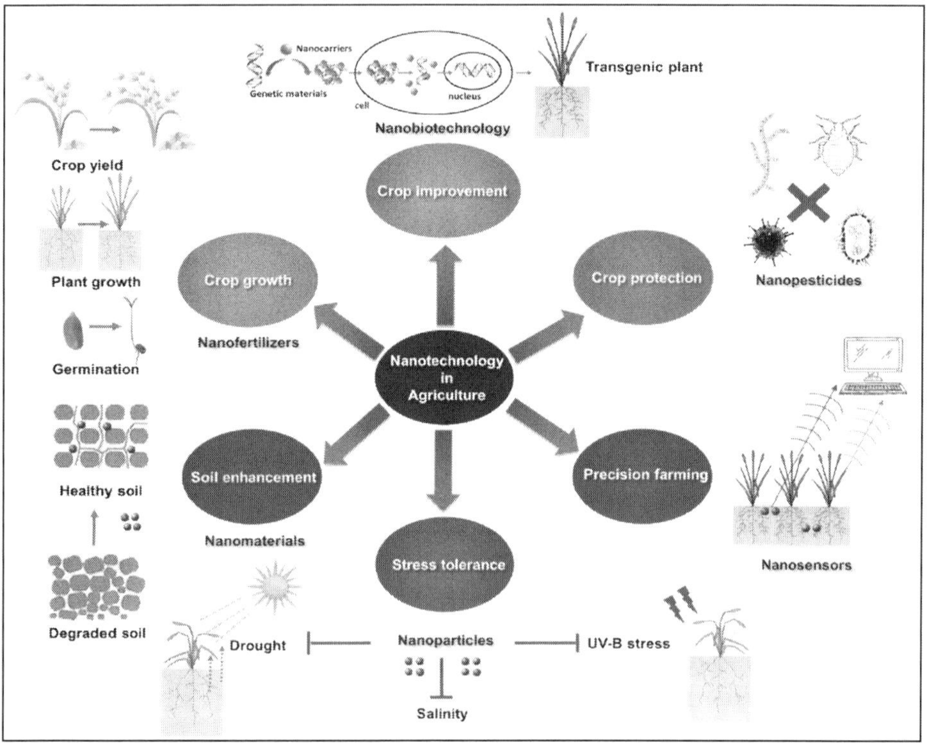

Figure 12.4. Application of Nanotechnology in Agriculture.

12.5 Nanofertilizers

Nanofertilizers are ntrient carriers of nanodimension capable of holding large quantity of nutrient ions due to their high surface area and release it slowly as per crop demand. **or** A nanofertilizer refers to a product in nanometer scale that delivers nutrients to crops. **or** Nanofertilizers are modified fertilizers synthesized by chemical, physical or biological methods using nanotechnology to improve their attributes and composition, which can enhance the productivity of crops.

Advantages

a) Release nutrients into the soil both gradually and in a controlled way

b) Controlled release of agrochemicals through site-targeted delivery

c) Reduction in toxicity
d) Enhanced nutrient utilization of delivered fertilizer
e) Enhance plants performance in terms of ultrahigh absorption, increase in production, rise in photosynthesis, and significant expansion in the leaves surface area
f) Preventing eutrophication
g) Preventing pollution of water resources
h) Enhance nutrient use efficiency

Difference between Conventional and Nanofertilizers

Nanofertilizers	*Conventional fertilizers*
High solubility	Low solubility due to large particle sizes
Improved dispersion of insoluble nutrients	Lower dispersion of insoluble nutrients
Reduce soil adsorption and fixation	High soil adsorption and fixation
Bioavailability high	Bioavailability low
Increased efficiency of nutrient uptake	Low efficiency of nutrient uptake
Release rate and pattern controlled	Excess release leading to toxicity
Extended effective duration	Used by the plant at the site and time of application and the rest is converted into insoluble from
Reduced loss of fertilizer nutrient	High loss rates

Nanoparticles can be loaded by nutrients most commonly through one of the following ways:

a) **Absorption on the nanoparticles:** *e.g.* chitosan nanoparticles suspensions containing N, P, and K fertilizers can be useful for agricultural application.

b) **Attachment on the nanoparticles mediated by ligands:** *e.g.* urea-modified hydroxyapatite (HA) nanoparticles are exploited for slow and sustained release of nitrogen over time with the crop growth. The large surface area of HA facilitates the large amount of urea attachment on the HA surface and the strong interaction between HA nanoparticles and urea contributes to the slow and controlled release of urea.

c) **Encapsulation in nanoparticulate polymeric shell:** *e.g.* Polymer-based mesoporous nanoparticles can also provide an efficient carrier system to agrochemical compounds.

d) **Entrapment in nanoparticles:** *e.g.* Mesoporous silica nanoparticles (150 nm) have been reported to entrap urea and to release it in a controlled manner in soil and water.

Application of Nanotechnology in Fertilizers

Treatment with TiO_2 nanoparticles on maize had a considerable effect on growth and increased the light absorption and photo energy transmission. Treatment with SiO_2 and TiO_2 nanoparticles on soybeans increased the activity of nitrate reductase and intensified the plant absorption capacity, making the use of water and fertilizer more efficient. A nano-organic iron-chelated fertilizer is proved to be environmentally sustainable. The positive effect from the uptake and penetration of ZnO_2 nanoparticles on tomato plants leaves supports its potential use as a future nano-fertilizer.

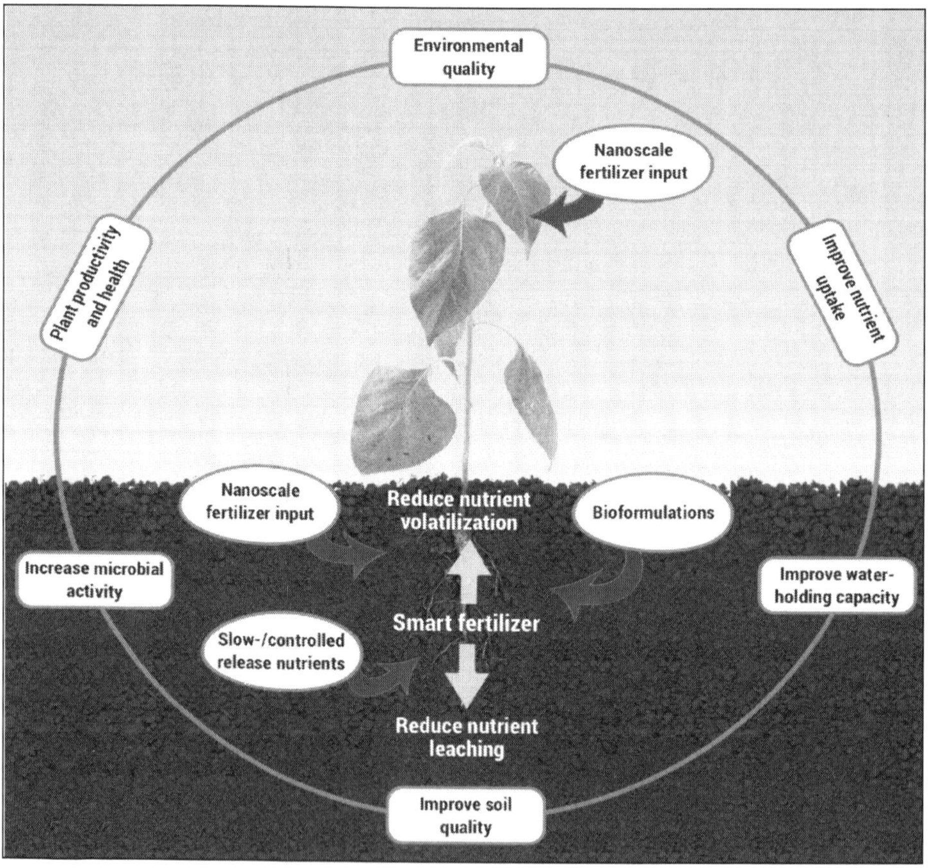

Figure 12.4. Schematic Diagram of Potential Smart Fertilizer Effect in the Soil Plant System.

Nano-fertilizers that ensure slow, targeted, efficient release have the potential to increase the efficiency of nutrient uptake. Engineered nano-particles are useful for mitigating the chronic problem of moisture retention in arid soils and enhancing crop production by increasing the availability of nutrients in the rhizosphere. Coating and binding of nano-particles help to regulate the release of nutrients from the fertilizer capsule. Application of a nano-composite consisting of nitrogen, phosphorus, potassium, micronutrients, mannose, and amino acids enhanced the uptake and use of nutrients by grain crops. Zn–Al layered double-hydroxide nano-composites have been employed for the controlled release of chemical compounds that act as plant growth regulators. Carbon nanotubes were found to penetrate tomato seeds and affect their germination and growth rates. Analytical methods indicated that the carbon nanotubes penetrated the thick seed coat and supported water uptake inside seeds. Multi-walled carbon nanoparticles (MWCNP) have been shown to promote seed germination and growth of tomato and enhance the growth of tobacco cells. The same phenomenon was observed in MWCNTs in mustard plants. Slow release of nanofertilizers and nanocomposites are excellent to soluble fertilizers. Nutrients are released at slower rate throughout the crop growth, plant are able to take up most of the nutrients without any waste. slow release of nutrients in the environment could be achieved by using zeolites that are a group of naturally occurring minerals having a honeycomb-like layered crystal structure. Its netwrok of interconnected tunnels and cages can be loaded with nitrogen and potassium, combined with other slowly dissolving ingredient containing phosphorous, calcium and a complete suite of minor and trace nutrients. Zeolite acts as a reservoir for nutrients that are slowly released "on demand". Fertilizers with sulphur nanocoating are useful slow release fertilizers stability of the coating reduced the rate of dissolutin of the fertiliser and allowed slow sustained release of sulphur coated fertilizer. In addition to sulphur nanocoating or encapsulation of urea and phosphate, their release will be beneficial to meet the soil and crop demands.

Smart delivery systems means a combination of specifically targeted, highly controlled, remotely regulated, and multifunctional characteristic to avoid biological barriers for successful targeting. The specific properties of nano-fertilizers, *i.e.* their high surface area, absorption capacity and controlled-release kinetics to targeted sites, attribute them as a smart delivery system. Smart fertilizers are becoming a reality through transformed formulation of conventional products using nanotechnology. The nanostructured formulation allows a fertilizer to intelligently control the release speed of nutrients in order to match the uptake pattern of a specific crop. It improves the solubility and dispersion of insoluble nutrients in the soil, reduces the soil absorption and fixation and increases the bioavailability and, hence, the nutrient uptake efficiency.

12.6 Nanopesticides

Nanopesticides are plant protection products where nanotechnology is employed to enhance the efficiency or reduce the enviroment impact of a pesticides active ingredient. or A nanopesticides is a plant protection product that contains components in the nanometer size range.

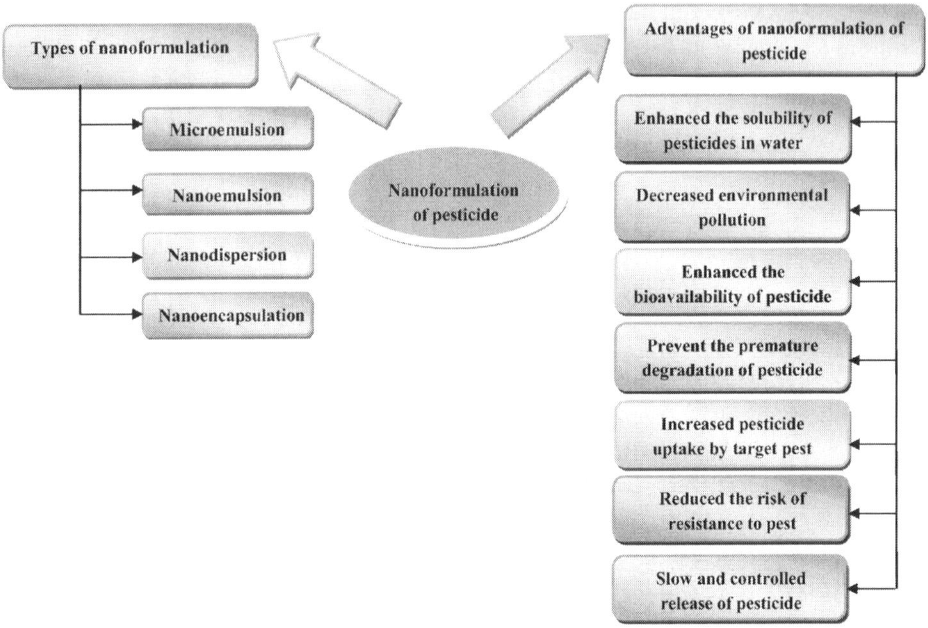

Figure 12.5. Nanopesticides: Formulation and Application in Agriculture.

Categories of Nano-Pesticides

Types	*Relation to Nano*	*Beneftis*
Nano-emulsions	Active ingredient in nano-scale oil droplets that float in an aqueous solution	Improved distribution and increased uptake by organisms
Nano- capsule	Active ingredient is packed in nanoscale capsules	Increase uptake and effectiveness
Nano-gel	Active ingredient is distributed in a gel which consist of nanoscale building blocks	Easier handling decreased evaporation of active ingredient
Nano-fibres	Active ingredient incorporated in a nanoscale fibred	Improved distributionslow release of active ingredient
Liposomes	Liposome's are in nano sized particles	Slow release of active ingredient
Lipid nanoparticles	Nanoparticles composed of lipids that carry the active ingredient	Slow release. protection from degradation
Inorganic nanoparticles	Inorganic nanoparticles as active ingredient	Improved distribution, high effectiveness of active ingredient against pests

Benefits of Nanopesticides

a) Controlled release
b) Increase efficiency towards pests
c) Increase the water solubility of active ingredient
d) Reduce the development of resistance of pests to pesticides
e) Lower risk to environment
f) Reduce the amount pesticide used
g) Improves the uptake and effectiveness of the pest control agent.

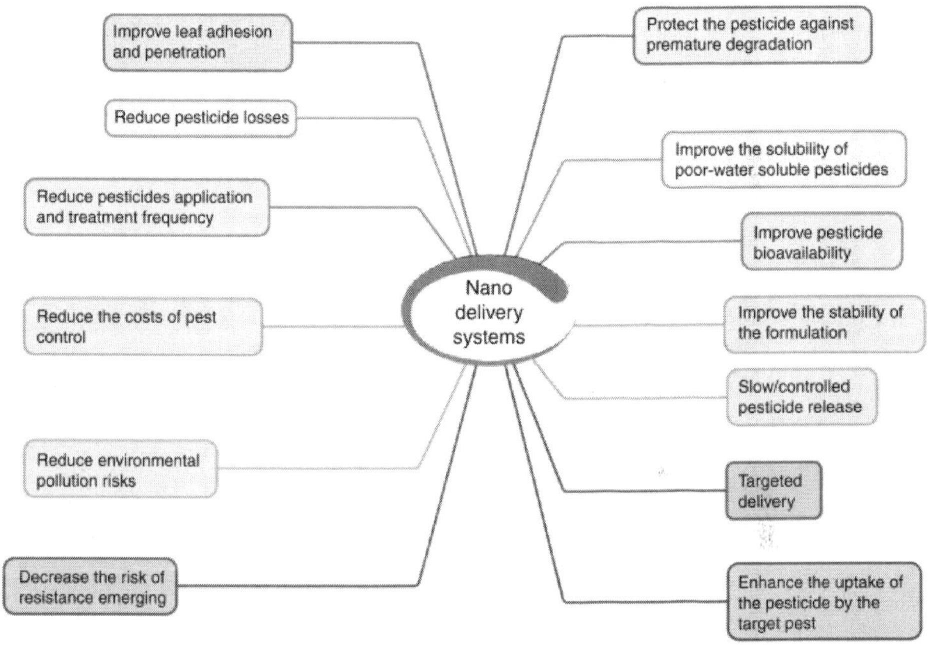

Figure 12.6. Advantage of Nanopesticides.

Application of Nanotechnology in Plant Protection

Spraying of pesticides involves either knapsacks that deliver large droplets wit splash loss causing spray drift. By using encapsulated or nanosized pesticides that will contribute to efficient spraying and reduction of spray drift and splash losses. Pesticide application in the field is settlement of formulation components in the spray tank and clogging of spray nozzles. Nanosized prevented spray tank filters from clogging did not required mixing and did not settle down in the spray tank due to the smaller sized particles. To protect the active ingredient from the adverse environmental conditions and to promote persistence, a nanotechnology approach, namely nanoencapsulation can be used to improve the insecticidal value. Nanoencapsulation of effective control of pests while preventing accumulation of residues in soil. In order to protect the active ingredient from degradation and to increase persistence. Nanopesticide will reduce the rate of application because the quantity of product actually being effective smaller than that applied.

Chapter 13

Important Objectives

1. In Remote Sensing **infrared rays** are used.
2. The main advantage of Nano fertilizer is **high surface area.**
3. Remote sensing techniques make use of the properties of **electromagnetic waves** emitted, reflected or diffracted by the sensed objects.
4. **Signature** helps to identify the objects on the earth surface.
5. The interaction of the electromagnetic radiation produced with a specific wave length to illuminate a target on the terrain for studying its scattered radiance is called **active remote sensing.**
6. The normal altitude of GPS satellite is about **20,200 km**
7. Orbital radius of GPS satellites is approximately **26, 600 km**
8. The infrared portion of EMR lies between **0.7 to 14 um**
9. Remote sensing uses **electromagnetic** waves in its procedure.
10. LANDSAT program Started at **1972**.
11. The primary colour of image is **BGR**.
12. **Roger L. Easton** is the inventor of GPS.
13. **Pisharoth Rama Pisharoty** is the father of Indian remote sensing.

14. **Roger F. Tomlinson** is the father of GIS.
15. **Red** color has the maximum wave length in visible spectrum.
16. Spatial Database is also called as **geodatabase.**
17. **Mr. Evelyn Pruitt** coined the term remote sensing.
18. DBMS stands for **Database management system.**
19. The information in GIS is entered and stored as **layers**.
20. **Vector and raster** are the two general data formats used in GIS.
21. **12 hrs** is the approximate time taken by the GPS for one complete orbit.
22. **24** is the number of GPS satellites used.
23. The GPS satellite launches begin from **1978.**
24. GIS stand for **geographical informatics system.**
25. GPS stand for **global positioning system.**
26. GPS was invented in **1970.**
27. **Al Myers** invented an on-the-go crop yield monitor.
28. **Pierre Robert** is the father of precision farming.
29. The prefix "nano" comes from a **Greek** word means **dwarf.**
30. National bureau of soil survey and land use planning is located at **Nagpur**.
31. Who first used the term nanotechnology. **Richard Feynman, 1959.**
32. STCR stands for **soil test crop response**
33. ICAR established the AICRP on STCR in **1967**.
34. The STCR concept was developed by **Ramamoorthy.**
35. The space application centre (SAC) is located at **Ahmedabad**.
36. The concept of precision agriculture first emerged in the **United States.**
37. **Janusz Bryzek** is the father of sensors.
38. **Biosensor** is an analytical device, used for the detection of a chemical substance, that combines a biological component with a physicochemical detector.

39. **John Deere** was the first to introduce precision farming technology using GPS location data from satellites.

40. 1 nm = $\mathbf{10^{-9}}$ **m.**

41. The size of nanoparticles is between **1 nm to 100 nm**.

42. "There is plenty of room at the bottom," This was stated by **Richard Feynmann**.

43. **Michael Faraday** was the first scientist to describe that substances having nanodimensions possess altogether different and unique properties.

44. The full form of STM is **scanning tunneling microscope**, invented by Gern Binnig and Heinrich Rohre in 1981.

45. **Sumio Tijima** prepared and explained nanotubes for the first time.

46. **Richard Smalley** conceptualised carbon nanotubes.

47. In remote sensing optical wave length from **0.3 to 15 ur** are used.

48. National Remote Sensing Agency is situated at **Hyderabad. NRSA**

49. The word remote sensing was coined by **Fischer** in 1960.

50. **GIS** is a planning tool for environmental studies.

51. **GIS** are powerful tool for the management and analysis of spatial data.

52. **Remote sensing** are powerful tools for collecting the classified spatial data.

53. Maharashtra Remote Sensing Application Centre (MRSAC), was set up by Government of Maharashtra in **1988** at **Nagpur**.

54. Indian Institute of Remote Sensing was established in **1966** at **Dehradun**, Uttarakhand.

55. Indian Institute of Surveying and mapping is located at Uppal, **Hyderabad**.

56. National Remote sensing Center is located at **Hyderabad**.

57. SSCM stands for **site specific crop management**.

58. DGPS stands for **Differential global positioning system.**

59. GNSS stands for **global navigation satellite system**.

60. A **satellite navigation or satnav system** is a system that uses satellites to provide autonomous geo-spatial positioning.

61. IRNSS stands for **Indian Regional Navigation Satellite System.** It is an autonomous regional satellite navigation system that provides accurate real-time positioning and timing services.

62. Precision agriculture is also known as **satellite** farming or site specific crop management (**SSCM**).

63. **Nucleation** is the phenomenon of initiation of formation of the first nano-crystals in solutions.

64. Gerd Binning and Heinrich Rochrer invented the scanning tunneling electron microscope (**STM**) and atomic force microscope (**AFM**).

65. The nanotechnology was first used in **1974** by **Noria Taniguchi.**

66. **Richard Feynman** is the father of nanotechnology.

67. **Training area** is a spectral features of areas of known land covers types are extracted from an image.

68. DSSAT stands for **Decision Support System for Agro-technology Transfer**.

69. APSIM stands for **Agricultural Production Systems Simulator**.

70. InfoCrop model developed by **Indian Agriculture Research Institute, New Delhi.**

71. A collection of non-redundant data which can be shared by different application systems is known as a **Database**.

72. CERES stands for **Crop Estimation through Resource and Environment Synthesis.**

73. AFM stands for **Atomic force microscope.**

74. **Interpretation** is the process of detection, identification, description and assessment of significant of an object and pattern imaged.

75. India's remote sensing programme under the **Indian Space Research Organization** (ISRO) started in 1988 with the IRS-1A.

76. **Pisharoth Rama Pisharoty** introduced remote sensing in India in 1970.

77. **Pisharoth Rama Pisharoty** was detected the coconut wilt disease using colour infrared aerial photographs.

78. Lidar (**light detection and ranging**) remote sensing is provides three-dimensional data through direct and indirect retrievals.

79. The GIS was developed in **Canada** during 1960 to analyze the Canada Land Inventory.

80. **Map** is a drawing of the whole or part of the surface of the Earth on a plane surface to a particular scale.

81. The art, science and technology dealing with the acquisition, storage, processing, production, presentation and dissemination of geoinformation is called **geoinformatics**.

82. Give some examples of vector file format. **Shapefiles (.shp), ArcSDE, DLG, KML, TIGER**.

83. Give some examples of raster file format. **JPEG2000, MrSID, netCDF, DRG, RPF, GIF.**

84. **Yield mapping** is the first step to determine the precise locations of the highest and lowest yield areas of the field and to analyze the factors causing yield variation.

85. SDSS stands for **Spatial Decision Support System.** It provides a framework for integrating; Crop modelling capabilities, Database management system, Policy makers expert knowledge and Map based outputs.

86. **GIS** is a computer system capable of assembling, storing, manipulating, and displaying geographically referenced information.

87. **Variable rate application management zones** are the natural variability of a field and are used to manage the VRA of inputs across the field.

88. **Prescription map** is an electronic data file containing specific information about input rates to be applied in every zone of a field.

89. **Crop scouting** is the process of precisely assessing pest pressure and crop performance to evaluate economic risk from pest infestations and disease, as well as to determine the potential effectiveness of pest and disease control interventions.

90. **Soil mapping** is the process of characterizing natural bodies of soils, classifying and grouping the defined soils into map units and capturing soil property information for interpreting and represent soil spatial distribution on a map.

91. **Soil map** is a geographical representation showing diversity of soil types, soil properties in the area of interest.

92. **Georeferencing** is the process of assigning coordinates to vectors or rasters so they can be oriented accurately on a model of the earth's surface.

93. **Geocoding** is the process of converting addresses into geographic coordinates which use to place markers on a map or position the map.

94. **Geoprocessing** is a framework and set of tools for processing geographic and related data.

95. **Variable rate applicator** is the process of applying different rates of crop inputs according to the variability within an agricultural field.

96. **Precision farming** is the management of inputs based on the needs of small sites of the fields with the use of variable rate application instead of conventional constant rate application.

97. **Simulation** is the process of building models and analyzing the system.

98. **Model** is a simplified representation of a system or a process.

99. **Geoinformatics** is the branch of sciences which uses the infrastructure of science to address the problems of geography, cartography, geosciences and related branches of science and engineering.

100. **Rasterization** is the process of converting data from vector to raster format.

101. **Vectorization** is the process of converting data from raster to vector format.

102. **Agro-informatics** is the application in agriculture with innovative ideas, techniques and scientific knowledge to expand the horizons of the Computer Science.

Glossary

Active sensing systems - sensing systems that generate a signal, bounce it off an object, and measure the reflected signal.

Actuator - a device used in variable-rate application that responds to controller signals to regulate the amount of material applied to a field.

Aerial photography - Photos taken from airplanes to assist growers to determine variations within an area of interest such as a field.

Agrogeoinformatics - A branch of geoinformatics, is the science and technology about handling digital agro-geoinformation.

Arc - A line described by an ordered sequence of points associated with vector data models. When a node joins two or more arcs and several arcs are linked together in a loop, they form an area or polygon.

ArcGIS - is a GIS software package produced by the Environmental Systems Research Institute (ESRI) which allows you to collect, store, manage, visualize, export, analyze and map geographic data.

Atomic clock - It is clock device that uses an electronic transition frequency in the microwave, optical or ultraviolet region of the electromagnetic spectrum of atoms as a frequency standard for its time keeping element.

Atomic force microscope - A scientific instrument that can generate images of nanoscale details on a physical surface by scanning a tiny, flexible ceramic or semiconductor probe just above the surface.

Attribute - A characteristics of a geographic feature, typically stored in tabular format and linked to the feature in a relational database.

Bushel - A unit of dry measure (1 cubic foot) for grain, fruit, *etc.*, equivalent to 8 gallons of liquid.

Carbon nanotube - Carbon molecule with a cylindrical shape.

Cartography - The art and science of the organization and communication of geographically related information such as a yield image into maps or charts.

Collect and Weigh - A method for determining crop yield, typically on a whole-field basis. Each truck or wagon load of grain is weighed as it leaves the field and the moisture content is determined by sampling the load.

Computer aided design (CAD) - Software with the capability of performing standard engineering drawings.

Control segment - the network of tracking stations that monitor and control GPS satellites

Coordinate system - Used in GPS/GNSS navigational systems to reference locations on Earth.

Crop scouting - Precise assessments of pest pressure and crop performance that can be tied to a specific location for better interpretation.

Database - A logical collection of files managed as unit. A GIS database includes data about both the position and the attributes of geographic features.

Data editing - It is the process of correcting errors in data input into a GIS.

Data input - The entry of information into a computer through the use of a keyboard, digitizer, scanner, or even entering data from already existing databases.

Data layer (in GIS) - A layer of information on a GIS map. A map can have many layers to present different types of information.

Database management system (DBMS) - A collection of software for organizing the information in a database that might contain routines for data input, verification, storage, retrieval and combination.

Decision support system (DSS) - a system usually computerized, dedicated to supporting decision regarding a specific problem or set of problems.

Deterministic model - a model for which there is only one possible answer for a given set of inputs.

Differential correction - correction of a GPS signal that is used to improve its accuracy using a stationary GPS receiver whose location is know.

Differential global positioning system (DGPS) - a method of using GPS which attains the position accuracy needed for precision farming through differential correction.

Digital image processing - Digital Image Processing is a collection of techniques for the manipulation of digital images by computers.

Dynamic model - A model in which time is the key variable while all other input variable remain constant.

Electromagnetic energy - Energy that is reflected or emitted from objects in the form of electrical and magnetic waves which can travel through space.

Electromagnetic spectrum - All wavelengths of electromagnetic energy including x-rays, ultraviolet rays, visible light, infrared light, microwaves and radio waves.

ESRI (Environmental Systems Research Institute): [software] Esri is the largest global supplier and developer of GIS software based on Redlands, California.

Feature - A geographic component of the earth's surface that has both spatial and attribute data associated with it.

Flow rate - Amount of seed distributed, expressed as a number, mass or volume of seed per unit of time.

Flow sensor - A sensor that measures the amount of material flowing through a conduit per unit of time.

Geocode - The process of identifying a location by one or more attributes from a base layer.

Geodesy - is a branch of applied mathematics and earth science of locating and assigning three dimensional points on Earth by measuring the shape of the Earth.

Geographic data - data that contains information about the spatial location (position) and the attribute being monitored. Also referred to as spatial data.

Geographic information system (GIS) - a computer-based system used to input, store, retrieve and analyze geographic data sets.

Geographic coordinate system - A reference system using latitude and longitude to define the locations of points on the surface of a sphere or spheroid.

Geoinformatics - It is the science and technologies which develops and uses information science and computer infrastructure to address the problems of geography, geosciences, related branches of engineering *etc.* Geoinformatics combines geospatial analysis and modeling, development of geospatial databases, information systems design, human-computer interaction and networking technologies.

Geometric correction - Correction to align measured ground control points in a remotely sensed mage with the ground control points on an established map of the area.

Georeferenced data - Spatial data that pertains to a location on the earth's surface.

Georeferencing - The process of associating non-spatial data such as crop yield values with geographic coordinate data to produce spatial data.

Geostationary satellite - Space vehicle in an orbit that keeps it over the same location on the earth at all times.

GPS (Global Positioning System) - A network of satellites controlled by the Defense Department that is designed to help ground based units determine their current location in latitude and longitude coordinates.

GPS antenna - The device that receives satellite signals from space.

GNSS (Global Navigation Satellite System) – Is the standard generic term for satellite navigation systems that provide geo-spatial positioning with global coverage using time signals transmitted from satellites.

Grain moisture content - Moisture content (MC) is the weight of water contained grain. The moisture content is generally reported on the wet basis meaning the total weight of the grain including the water.

Graphical user interface (GUI) - Computer interface such as MS Windows, that is used to provide a user friendly interface between a user and the computer.

Greenwich Meridian is passes through the Royal Observatory in Greenwich, England and is defined as a 0° line of longitude.

Grid - A pattern of regularly spaced horizontal and vertical lines forming square zones on a map used as a reference for establishing points.

Grid soil sampling - The collection of samples from small, uniform-sized cells based on a systematic grid laid out across a field. Grid location in the field is used to develop a field map for the attribute measured.

Ground control points - stationary objects/areas on the earth's surface that provide georeferenced points in a remote sensing image/aerial photograph.

Ground reference data - The field collection of data that is used in the interpretation of information gathered from other sources such as a yield image or a remotely sensed image.

Hardware - The various physical components of an information processing system such as a computer, view screen, plotters and printers.

Hyperspectral sensors - Sensors capable of measuring hundreds of individual wavelengths simultaneously.

Image - A raster based representation or description of a scene, typically produced by an optical or electronic device.

Image classification - It is the process of sorting all the pixels in an image into a finite number of individual classes.

Instantaneous yield monitor - A yield monitor that continuously measures and records crop yields on-the-go.

Ionosphere - A blanket of electrically charged particles 80 to 400 km above the earth

Landscape - A collective term for all the natural features (such as fields, hills, forests, water, *etc.*) that distinguish one part of the earth's surface from another part.

LANDSAT (LAND SATellite) - a series of U.S. satellites used to study the earth's surface using remote sensing techniques.

Latitude/Longitude - A coordinate system that is used to identify positions on earth. Latitude is the north to south position. Longitude is the east to west position. Locations are described in units of degrees, minutes and seconds.

Layer - A logical separation of mapped information representing common data.

Lidar - (light detection and ranging) is an optical remote-sensing technique that uses laser light to densely sample the surface of the earth, producing highly accurate x,y,z measurements.

Line - Lines represent geographic feature too narrow to be displayed as an area at a given scale, such as contour, street, streams.

Load cell - A device that converts the effect of a force or weight into an electrical signal.

Management zone - Management zones are created by subdividing a field into 10-20 acre areas with similar characteristics. Yield maps, soil texture maps, elevation data, EC data, sensor data and farmer knowledge can be used to create management zones in GIS software.

Map - A graphic representation of geographically distributed phenomena.

Mass flow sensor - Is a sensor that measures grain flow in a yield monitor system.

Modelling - is the application of methods to analyse complex, real word problems in order to make prediction about what might happen with various actions.

Moisture sensor - Is a sensor that measures grain moisture in a yield monitor system.

Multispectral - A data product in which the data are binned into as few as two, or as many as approximately fifteen.

Multispectral image - Image optically acquired in more than one spectral or wavelength interval. Each individual image is usually of the same physical area and scale but of a different spectral band.

Nanofertilizers - It is any product that is made with nanoparticles or uses nanotechnology to improve nutrient efficiency.

Nanofrication - General terms for methods to create, assemble or form nanoscale structure.

Nanometers - A distance measure equal to one millionth of 1 mmmm.

Nonoparticles - A microscopic particle of matter that is measured on the nanoscale, usually one that measures less than 100 nanometers.

Nanopesticides - It is any product that is made with nanoparticles or uses nanotechnology to improve chemical efficiency.

Nanoscale - The size range roughly 1 to 100 nanometers.

Nanoscience - is the study of unique properties of matter at the nanoscale.

Node - is an important point along a line feature where two lines intersect.

Nanosensor - is a device for sensing radiation, forces, chemicals or biological agents, in which some portion of a device operates at the nanoscale.

Nanotube - is a Hollow, cylindrical structures, with a diameter usually less than 5 nanometers.

Passive sensing systems - Sensing systems that measure naturally-emitted and reflected signals.

PCMCIA - Personal Computer Memory Card International Association.

Photosensor - A sensor that is used to detect light.

Pixel - An abbreviation of "picture element" - the smallest area or element of an image map. A pixel is represented in a remotely-sensed image as a rectangular cell in an array of data values and contains a data value that represents a measurement of some real-world feature.

Point - A single coordinate or pixel used to describe the location of an object too small to be represented as a polygon.

Polygon - An area enclosed by a line describing spatial elements, such as a similar crop yields range, land use, or soil type.

Positioning system - A general system for identifying and recording, often electronically, the location of an object or person.

Precision farming - managing crop production inputs (seed, fertilizer, lime, pesticides, *etc.*) on a site-specific basis to increase profits, reduce waste and maintain environmental quality.

Prescription - refers to the map created in an AgGIS which assigns product application rates for variable rate applications.

Pseudorange - An estimation of the true distance (range) from a GPS receiver to a satellite.

Radar (Radio Detection and Ranging) - A method of determining the position or velocity of an object by bouncing high frequency signals off the object and measuring the reflected signal.

Radiometric correction - Correction to reduce remotely-sensed image distortion from variations in radiation levels at the time of sensing.

Radiometric data - Information on crop or soil conditions obtained by sensors detecting visible and invisible light reflected from surfaces.

Radiametry - The measurement of electromagnetic radiation, including visible and invisible light waves.

Raster format - Format for storing GIS spatial data in which the data are stored in cells which are addressed by the row and column of the cell.

Rasterization - the process of converting data from vector to raster format.

Rate controller - An electronic device that varies the amount of chemical/plant nutrient applied to a given area.

Receiver (in GPS hardware) – A computer-radio device that receives satellite information by way of radio waves to determine the position of its antenna relative to the earth's surface.

Remote sensing - the act of monitoring an object without direct contact between the sensor and object.

Resolution - the smallest feature that can be mapped and measured.

Satellite - A communications vehicle orbiting the Earth.

Satellite constellation - A system of 24 satellites that is owned by the US Department of Defense (DOD) that can determine location to within inches. There are usually at least 4 of these satellites that are in view 24 hours a day.

Satellite image - Graphical image taken of the Earth surface using electromagnetic sensors.

Scale - The ratio between the distance on a ma p, chart, or photograph and the corresponding distance on the ground.

Scanner - Raster input device used to convert maps and other analogue data into digital data using the raster data model.

Scanning probe lithography - A technique in which an atomic force microscope or scanning tunneling microscope scratches, indents, or heats to produce nanoscale features on a surface.

Scanning tunneling microscope - A scientific instrument that can make images of nanoscale details on an electrically conductive surface by moving a sharp metal probe very close to that surface, passing a low-voltage electric current across it, and measuring tiny fluctuations in the current as the probe is scanned across the surface.

Sensor based variable rate application system - A system that a adjusts product application rate on-the-go based on information received from real time sensors.

Sensor technologies - Sensor technology refers to on-the-go optical sensors used to measure crop status.

Simulation - It is a technique for studying real world dynamical systems by imitating their behaviour using a mathematical model of the system implemented on a digital computer.

Site specific crop management (SSCM) - The use of yield maps, grid sampling and other precision tools to manage the variability of soil and crop parameters and aid decisions on production inputs.

Site specific farming - it is managing areas within field, rather than using the same management on the entire field.

Site specific management - it is used to detect and measure the difference within fields, record these difference at specific location and the use this information to guide changes in management.

Soil map - A plan of soil characteristics such as texture or fertility, created from a set of samples taken across an area at different depths.

Soil mapping - The production of a map of a field showing differences in soil characteristics.

Soil sampling - taking small amounts of soil from a site to analyse composition, nutrient content or condition.

Soil series - A grouping of soils which have developed from a particular kind of parent material and which are similar in all characteristics except texture of the surface layer.

Soil survey - The systematic examination, description, classification, and mapping of soils in an area.

Soil testing - The analysis of soil samples to establish composition or nutrient content.

Soil variability - Differences in soil type and fertility across an area, as a result of previous cropping patterns, fertiliser use, underlying soil texture or compaction.

Space segment - The portion of G PS consisti ng of NAVSTAR satellites orbiting the earth at 20,200 km.

Spatial data - Data that contains information about the spatial location (position) and the attribute being monitored such as yield, soil properties, plant variables, seed population, *etc.*

Speed sensors - Sensors that measure the rotational speed of a shaft or the reflection of radio or sound waves off the ground to determine machine speed.

Spectral band - A well defined continuous wavelength range in the spectrum of reflected or radiated electromagnetic energy.

Stochastic model - A model that recognizes that there could be a range of possible outcomes for a given set of inputs.

Systematic sampling - It involves selecting samples using a regular and ordered sampling framework.

Thematic map - Result from remote sensing data classification that produces categorization of all pixels in a digital image into classes or themes.

Theme - A theme usually represents a single geographic layer, such as soil, roads or land use.

User segment - The portion of GPS consisting of receivers used by civilians and the military for determining the position of a person or object.

Variable rate application (VRA) - Adjustment of the amount of cropping inputs such as seed, fertilizer, and pesticides to match conditions in a field.

Variable eate technology - GPS and precise placement technology that uses an "application guidance" map to direct the application of a product to a specific, identifiable location within a field.

Vector - A coordinate based data model that represents geographic feature as points, lines and polygons.

Vector format - A format for storing and displaying GIS spatial data in which the data is stored as points, lines, or areas that create the terrain or map objects.

Vectorization - The process of converting data from raster to vector format.

Vertex - It is a point that specifies a position on a line for arcs, polylines and polygons.

Yield maps - The process of using GPS and yield monitoring data to show the variation in yield across a field.

Yield monitor - A device on a combine harvester that measures crop yield against field position.

Yield monitoring - Yield monitoring allows growers to determine higher and lower yielding areas of the field.

Yield variation - It is differences in yield across a field.

World Wide Web - Interconnected global network of computers and the software interface used to access and exchange digital information and multimedia.

Zone management - The information-based division of large areas into smaller areas for site specific management applications.

References

Aronoff, S. (1989). Geographical information systems: A Management Perspective, WDL Publications, Ottawa, Canada.

Arslan, S. and Colvin, T. S. (2002). Grain yield mapping: yield sensing, yield reconstruction and errors. Precision Agriculture. 3: 135-154.

Barocco, R., Lee, W. S. and Hortman, G. (2017). Yield mapping hardware components for grains and cotton using on-the-go monitoring systems. Department of Agricultural and Biological Engineering, UF/IFAS Extension. 1-10. http: //edis.ifas.ufl.edu.

Bongiovanni, R. and Debore, J. L. (2004). Precision agriculture and sustainability. Precision Agriculture, 5. 359-387.

Chandel, N. S., Mehta, C. R., Tewari, V. K. and Nare, B. (2016). Digital map based site specific granular fertilizer application system. Current Science, 111(7): 1208-1213.

Das, D. K. (2018). Introductory soil sciences. Kalyani Publishers, Ludhiana, New Delhi.

Dodd, R. B., Han, Y. J., Khalilian, A. and Keskin, M. (1999). Farm mechanization in USA for environment friendly agriculture - Environmentally friendly agriculture: Is it possible? Proceedings of the international symposium on farm mechanization for environment - friendly agriculture. Seoul, South Korea. pp. 5-26.

Dwivedi, A., Naresh, R. K., Kumar, R., Yadav, R. S. and Kumar, R. (2017). Precision agriculture. Agri-horticultural, technological innovations. Parmar publishers and distributors, Dhanbad, Jharkhand, 5. 83-105.

Four decades of STCR research - crop wise recommendations. AICRP on soil test crop response correlation, Indian Institute of soil sciences, Bhopal.

Grisso, R. D., Alley, M. M., Thomason, W. E., Holshouser, D. L., and Roberson, G. T. (2011). Precision farming tools: Variable-rate application. Virginia Cooperative Extension publication 442-505.

Guo, T., Nikolaev, P., Thess, A., Colbert, D. T. and Smalley, R. E. (1995). Catalytic growth of single-walled nanotubes by laser vaporization. Chemical Physics Letters. 243: 49-54.

Hakkim, V. M. A., Joseph, E. A., Gokul, A. J. A. and Mufeedha, K. (2016). Precision farming: The future of Indian agriculture. Journal of Applied Biology and Biotechnology. 4(6): 68-72.

Hoque, Z. M. (2016). Basic concept of GPS and its applications. Journal of Humanities and Social Science. 21(3): 31-37.

https: //cropwatch.unl.edu/ssm/mapping

https: //en.wikipedia.org/wiki/International_Cartographic_Association

https: //en.wikipedia.org/wiki/American_Society_for_Photogrammetry_and_Remote_Sensing

https: //en.wikipedia.org/wiki/Precision_agriculture

https: //en.wikipedia.org/wiki/Characterization_of_nanoparticles

https: //m.farms.com/precision-agriculture/yield-monitoring

www.nationalgeographic.org

http: //precisionagriculture.unl.edu

https: //www.asprs.org/

https: //www.bio-fit.eu/q1/lo4-nano-fertilizers-and-genetically-engineered-microbes

https: //www.nanopartikel.info/en/nanoinfo/nanopesticides-nano-in-plant-protection products

https: //www.academia.edu/

http: //www.apsim.info/apsim/Documentation

http: //www.awgfarms.com/Yield_Monitors_and_Yield_Mapping.html.htm

http: //www.aboutcivil.org/components-of-gps-working-mechanism.html

http: //www.fao.org/nr/water/infores_databases_aquacrop.html

Janipella, R., Gupta, V. and Moharir, R. V. (2019). Application of geographic information system in energy utilization. Current Developments in Biotechnology and Bioengineering.pp.143-161.https: doi.org/10.1016/B978-0-444-64083-3.00008.7.

Jones, J. W., Hoogenboom, G., Porter, C. H., Boote, K. J., Batchelor, W. D., Hunt, L. A., Wilkens, P. W., Singh, U., Gijsman, A. J. and Ritchie, J. T. (2003). The DSSAT cropping system model. European Journal of Agronomy. 18 (3-4): 235-265.

Kumar, N. S., Kasturi Bai, K.V., Rajagopal, V., Aggarwal, P. K. (2008). Simulating coconut growth, development and yield using InfoCrop-coconut model. Tree Physiol. 28, 1049–1058.

Kant, K., Telkar, S. G., Gogoi, M. and Kumar, D. (2017). Crop simulation model. Bio-molecule reports- An International eNewsletter. ISSN: 2225-8758: 1-4.

Keskin, M, Han, Y. J. and Dodd, R. B. (1999). A review of yield monitoring instrumentation applied to the combine harvesters for precision agriculture purposes. 7th International Congress on Agricultural Mechanization and Energy. Adana, Turkey. pp. 426-431.

Khanna, P. Kaur, A. and Goyal, D. (2019). Algae based metallic nanoparticles: Synthesis, characterization and applications. Journal of Microbiological Methods. 163: 1-24.

Kosamkar, P. K, and V. Y. Kulkarni, V. Y. (2019). Agriculture crop simulation models using computational intelligence. International Journal of Computer Engineering and Technology. 10 (3): 134-140

Kushwaha, S. P. S. (2008). Geoinformatics - technology and applications. Zoological Survey of India, Kolkata. ENVIS Newsletter 14 (1 and 2): 2-13.

Mankari, M. P., Kodge, B. G., Kulkarni, M. J. and Nagargoje, A. U. (2010). Fundamentals of geoinformatics and its applications in geography. Geoscience Research. 1(1): 1- 6.

Mehta, A. and Masdekar, M. (2018). Precision agriculture- A modern approach to smart farming. International Journal of Scientific and Engineering Research. 9(2): 23-26.

Mourdikoudis, S., Pallares, R. M. and Thanh, N. T. K. (2018). Characterization techniques for nanoparticles: comparison and complementarity upon studying nanoparticle properties. The Royal Society of Chemistry, Nanoscale, 10, 12871–12934.

Murthy, V. R. K (2002). Crop growth modeling and its applications in agricultural meteorology. Satellite Remote Sensing and GIS Applications in Agricultural Meteorology. pp. 235-261.

Nagula, S. and Usha, P. B. (2016). Application of nanotechnology in soil and plant system with special referenceto manofertilizers. Advances in Life Sciences 5(14): 5544-5548.

Patra, J. K. and Baek, K. H. (2014). Green nanobiotechnology: factors affecting synthesis and characterization techniques. Journal of Nanomaterials. http: // dx.doi.org/10.1155/2014/417305

Patricia Oteng-Darko, Yeboah, S., Addy, S. N. T., Amponsah, S. and E. Owusu Danquah (2013). Crop modeling: A tool for agricultural research- A review. Journal of Agricultural Research and Development. 2(1): 1-6.

Raju, P. L. N. (2004). Fundamentals of geographical information system. Satellite Remote Sensing and GIS Application in Agricultural Meteorology. pp. 103-120.

Raju, P. L. N. (2004). Fundamentals of GPS. Satellite Remote Sensing and GIS Applications in Agricultural Meteorology. pp. 121-150

Rakshit, A., Raha, R. and De, N. (2015). Manures, fertilizers and pesticides. CBS publishers and distributors pvt. ltd., Daryaganj, New Delhi.

Reddy, S. R. (2019). Principles of agronomy. Kalyani Publishers, Ludhiana, New Delhi.

Reddy, S. R. (2019). Geoinformatics and nanotechnology for precision farming. Kalyani Publishers, Ludhiana, New Delhi.

Sah, H. N. (2017). Basic concepts of the global positioning system and its future applications. Journal of Emerging Technologies and Innovative Research. 4(11): 919-924.

Sahu, D. D. and Solanki, R. M. (2012). Remote sensing techniques in agriculture. AGROBIOS, India.

Saravanan, P., Gopalan, R. and Chandrasekaran, V. (2008). Synthesis and characterisation of nanomaterials. Defence Science Journal. 58 (4): 504-516.

Sahoo, R. N. (2010). Geoinformatics for precision agriculture. Lecture delivered in FAI Workshop on Fertiliser Reform through ICT at Kufri Holiday Resort, Kufri, Shimla.

Shanmugapriya, P, Rathika, S., Ramesh, T. and Janaki, P. (2019). Applications of Remote Sensing in Agriculture - A Review. International Journal of current microbiology and applied Sciences, 8(1): 2270-2283.

Sharma, R., Aashish Kumar, A., Kapil Malik, K., Abhilash and Harender (2017). Application of remote sensing in agriculture. Mahima Research Foundation and Social Welfare,19: 141-146.

Sinha, N. K., Mohanty, M., Somasundaram, J., Shinogi, K. C., Hoti, K. M. and Chaudhary, R. S. (2018). Application of remote sensing in agriculture. Harit Dhara, 1(1): 15-16.

Srivastava, R. (2012). Synthesis and characterization techniques of nanomaterials. International Journal of Green Nanotechnology, 4: 17–27.

Stockle, C. O. Donatelli, M. and Nelson, R. (2003). CropSyst, a cropping systems simulation model. European Journal of Agronomy. 18 (3–4): 289-307.

Tangwongkit, R., Salokhe, V. and Jayasuriya, H. (2006). Development of a tractor mounted real time, variable rate herbicide applicator for sugarcane planting. Agricultural Engineering International: the CIGR Ejournal. Manuscript PM 06 009. Vol. VIII.

Thakkar, K. N., Snehit, S., Mhatre, M. S., Rasesh Y. and Parikh, M. S. (2009). Biological synthesis of metallic nanoparticles. ww.sciencedirect.com

Vanuytrecht, E. Raes, D., Steduto, P., Hsiao, T. C., Fereres, E., Heng, L. K., Vila, M. G. and Moreno, P. M. (2014). AquaCrop: FAO's crop water productivity and yield response model. Environmental Modelling and Software. 62: 351-360.

Varshneya, M. C. and Pillai, P. B (2018). Text book of agricultural meteorology. Indian Council of Agricultural Research, New Delhi.

Wollenhaupt, N. C. and Wolkowski, R. P. (1994). Grid soil sampling. Better Crops/ Fall. 78(4): 6-9.

www.forestrymaps.alaska.gov.

Zakka, U., Lawal, O. and Luke ChinaruNwosu, L. (2019). Leveraging on agricultural entomology in precision farming for sustainable agriculture and food security. Canadian Journal of Agriculture and Crops, 4(2): 173-187.

Index

W

Y

Z